MW01697501

Since changing the climat[e]
with the gravest implicati[ons]
produces all kinds of emotions—conscious and unconscious—in those who apprehend it. This fine book helps grapple with those feelings, in thoughtful and empowering ways.

—**BILL McKIBBEN,** AUTHOR OF *HERE COMES THE SUN: A LAST CHANCE FOR THE CLIMATE AND A FRESH CHANCE FOR CIVILIZATION*

At a moment when many feel overwhelmed or paralyzed, this book offers a welcome path that honors our grief while steadily guiding us back into agency, connection, and meaningful action.

—**CHRISTIANA FIGUERES,** HEAD OF THE UN CLIMATE CHANGE CONVENTION 2010-2016

This book is rich with practical wisdom for all climate feelers, thinkers, and doers. Leslie Davenport believes in us. *Transforming Climate Anxiety* will leave you saying, "I do too."

—**KATHARINE K. WILKINSON, DPhil,** AUTHOR OF *CLIMATE WAYFINDING: HEALING OURSELVES AND THE PLANET WE CALL HOME*; COEDITOR OF *ALL WE CAN SAVE: TRUTH, COURAGE, AND SOLUTIONS FOR THE CLIMATE CRISIS*

Leslie Davenport is a guiding light for me and many others who are struggling to process the climate crisis and to parent through it. This book is another wise and welcome offering that will be an invaluable resource for individuals and communities.

—**ANYA KAMENETZ,** AUTHOR, *THE GOLDEN HOUR* NEWSLETTER AND *FALLING IN LOVE WITH THE WORLD AGAIN: FINDING YOUR WAY WHEN EVERYTHING FEELS BROKEN*

In *Transforming Climate Anxiety*, Leslie Davenport helps us stay with the emotional weight of the climate crisis without turning away. Grounded in climate psychology, this workbook brings together clear insights and practical exercises to help readers work through difficult emotions while staying connected to purpose, community, and action. This is an important book for anyone trying to remain engaged with the climate crisis, even when the realities feel overwhelming.

–**IMOGEN R. HERRICK, PhD,** ASSISTANT PROFESSOR OF STEM EDUCATION, DEPARTMENT OF CURRICULUM AND TEACHING, UNIVERSITY OF KANSAS, LAWRENCE, KS, UNITED STATES

With grace, elegance, and expansive compassion, Leslie Davenport takes us on a very compelling journey. Through accessible and gentle exercises, she helps us move from very personal internal distress to meaning, expression, and external collective action on behalf of our community, future generations, and the wider planetary ecosystem we call home. For anyone who loves living on planet Earth, the journey offered in this workbook is subtle, accumulative, and unmistakably powerful—expanding step-by-step our capacity to stay in the uncertain with curiosity, connection, and care.

–**REBECCA B. WESTON, JD, LCSW,** CO-EXECUTIVE DIRECTOR, CLIMATE PSYCHOLOGY ALLIANCE OF NORTH AMERICA

Leslie Davenport's workbook pulls no punches in confronting the fossil fuel industry and its ongoing role in driving the climate crisis. With clarity and urgency, she lays bare the scale of environmental damage and the systemic forces that have allowed it to persist. Davenport argues that the industry's complicity is no longer debatable—it is documented, measurable, and deeply entrenched. Yet, rather than leaving readers in despair, she challenges us to reckon with this reality and consider what accountability, transformation, and collective action must look like in the face of an escalating global emergency.

—**BARBARA EASTERLIN, PhD,** PRESIDENT OF EXECUTIVE COMMITTEE, CLIMATE PSYCHOLOGY ALLIANCE OF NORTH AMERICA

Transforming Climate Anxiety: A book that is grounded in care, resilience, and a world where we can nourish our relationships with ourselves.

—**ISAIAS HERNANDEZ,** ENVIRONMENTALIST, EDUCATOR, QUEERBROWNVEGAN

This well-researched and informative book, based on the recognition that climate change is a problem not only for environmental health but also for mental health, gives practical exercises to help people who are feeling worried and anxious to understand and cope with their responses. Davenport makes the case that both personal and ecological resilience will benefit when we acknowledge and strengthen connections to wider social and ecological networks.

—**SUSAN CLAYTON, PhD,** WHITMORE-WILLIAMS PROFESSOR OF PSYCHOLOGY, PSYCHOLOGY DEPARTMENT, THE COLLEGE OF WOOSTER, WOOSTER, OH, UNITED STATES

At once practical and deeply resonant, this workbook is a timely and much-needed contribution that will be of benefit to anyone feeling the pain of climate anxiety and grief. As we navigate our climate emotions during these challenging times, we are fortunate to have Davenport's knowledgeable and wise presence guiding us along the pathway, through our feelings and toward the connection and meaning at the heart of resilience.

—**LAURA CARTER ROBINSON, PsyD,** ECO- AND CLINICAL PSYCHOLOGIST; AMERICAN PSYCHOLOGICAL ASSOCIATION'S CLIMATE CHANGE PSYCHOLOGY ADVISORY GROUP

Climate change evokes emotions in all of us, albeit in different forms. Leslie Davenport has been a forerunner in developing constructive methods for engaging with climate emotions, and this book offers people both succinct information and easy-to-use practices—warmly recommended!

—**PANU PIHKALA, PhD,** LEADING INTERDISCIPLINARY SCHOLAR OF ECO-EMOTIONS; ADJUNCT PROFESSOR, THEOLOGY AND HELSINKI INSTITUTE OF SUSTAINABILITY SCIENCE HELSUS, UNIVERSITY OF HELSINKI, HELSINKI, FINLAND

This book is for everyone who cares deeply about climate change and needs practical skills to resist overwhelm and burnout. It shows you how to do the internal emotional work, connect it to the external community work, and help build a better world. I promise, you will come back to it over and over again.

—**JADE S. SASSER, PhD,** ASSOCIATE PROFESSOR, DEPARTMENT OF SOCIETY, ENVIRONMENT, AND HEALTH EQUITY, UNIVERSITY OF CALIFORNIA, RIVERSIDE, CA, UNITED STATES; AUTHOR, *CLIMATE ANXIETY AND THE KID QUESTION: DECIDING WHETHER TO HAVE CHILDREN IN AN UNCERTAIN FUTURE*

Leslie Davenport's book is a must-read for anyone experiencing the emotional toll of climate change. Through thoughtful activities and teachings, this is a straightforward guide to taking meaningful action that also fosters essential self-care. Action and well-being are intertwined, and Leslie's accessible and impactful book helps build these essential skills.

—**SARAH NEWMAN,** FOUNDER AND EXECUTIVE DIRECTOR, CLIMATE MENTAL HEALTH NETWORK

Climate psychology is genuinely complex work. It asks us to hold difficult emotions alongside the need to stay engaged, and keep our hearts and minds open. Leslie Davenport navigates that terrain with steadiness and skill. She has been one of the most trusted voices in the emerging field of climate psychology, and this book makes plain why. Reading this author's work feels like sitting with a compassionate guide, drawing on her clinical wisdom and warm intelligence. The result is a grounded, practical resource that so many of us in this field have been needing. It is an important contribution, and it confirms Leslie's place as one of climate psychology's essential voices.

—**RENÉE LERTZMAN, PHD,** AUTHOR OF *ENVIRONMENTAL MELANCHOLIA: PSYCHOANALYTIC DIMENSIONS OF ENGAGEMENT*

Put this book on your bookshelf; it is one of the good books about psychology in the face of the climate crisis, climate anxiety, and how to navigate these challenging times. Practical, readable, relatable, thoughtful, and kind, it gives us a deeper understanding about how to find a way through these times with heart and soul. Well worth a read; it will be a good companion to go back to again and again.

—**CAROLINE HICKMAN,** CLIMATE AWARE PSYCHOTHERAPIST AND RESEARCHER

Transforming Climate Anxiety

Transforming Climate Anxiety

A WORKBOOK FOR COURAGE, CLARITY, AND COLLECTIVE ACTION

LESLIE DAVENPORT

AMERICAN PSYCHOLOGICAL ASSOCIATION

The opinions and statements published are those of the Author, and do not necessarily represent the policies of the American Psychological Association. The information contained in this work does not constitute personalized therapeutic advice. Users seeking medical advice, diagnoses, or treatment should consult a medical professional or health care provider. The Author has worked to ensure that all information in this book is accurate at the time of publication and consistent with general mental health care standards.

Published by
American Psychological Association
750 First Street, NE
Washington, DC 20002
https://www.apa.org

Order Department
https://www.apa.org/pubs/books
order@apa.org

Typeset in Sabon by Circle Graphics, Inc., Reisterstown, MD

Cover Designer: Mark Karis
Back Cover Photograph: Photo courtesy of Patricia Kay Spinks

Library of Congress Cataloging-in-Publication Data

Names: Davenport, Leslie (Climate psychology educator and consultant) author
Title: Transforming climate anxiety : a workbook for courage, clarity, and collective action / by Leslie Davenport.
Description: Washington, DC : American Psychological Association, [2026] | Includes bibliographical references and index.
Identifiers: LCCN 2026021367 (print) | LCCN 2026021368 (ebook) | ISBN 9781433847967 paperback | ISBN 9781433847974 epub | ISBN 9781433851742 pdf
Subjects: LCSH: Climatic changes--Psychological aspects | Environmental psychology | Anxiety--Treatment | BISAC: SELF-HELP / Anxieties & Phobias | NATURE / Environmental Conservation & Protection | LCGFT: Handbooks and manuals
Classification: LCC BF353.5.C55 D383 2026 (print) | LCC BF353.5.C55 (ebook)
LC record available at https://lccn.loc.gov/2026021367
LC ebook record available at https://lccn.loc.gov/2026021368

https://doi.org/10.1037/0000511-000

Printed in the United States of America

10 9 8 7 6 5 4 3 2 1

CONTENTS

Transforming Climate Anxiety

INTRODUCTION

COLLECTIVE SOLUTIONS TO A COLLECTIVE PROBLEM

We are confronted every day with news about floods, droughts, and raging wildfires. An August 2023 report from scientists on the Intergovernmental Panel on Climate Change warned us again that a disturbing level of severe ecological degradation from climate change is occurring much sooner than anticipated.

Researchers use the term *eco-anxiety* to describe the worry, fear, and grief associated with environmental change. It is easy in these times to fall into eco-anxiety or despair, but those feelings mean that we are paying attention and care about what is happening. As climate impacts increase, so too does climate-triggered emotional distress. In November 2023, Google reported that searches for "climate anxiety" or "eco-anxiety" had increased 4,590% from 2018 to 2023 (Moench, 2023).

However, even with more of us becoming aware of the climate crisis, business as usual persists. Why do we continue living the way we do given the overwhelming evidence of the damage our lifestyles are causing to ourselves and the entire biosphere? What can we do to make a difference when we feel so small in the face of such a complex crisis? How can we work with the waves of overwhelming feelings?

Transforming Climate Anxiety is for anyone who feels overwhelmed, anxious, stuck, burned out, or heartbroken about the climate crisis. Whether you are a parent, activist, scientist, journalist, student, or educator—if your heart is awake to what is unfolding on our planet and you are ready to make a difference, you belong here. Through clear guidance, reflective prompts, somatic practices, and grounding psychological tools, you will learn how to regulate the nervous system, expand your emotional capacity, and transform distress into purposeful engagement. The goal is not to suppress emotions but rather to understand them as a sign of our deep care. This book will support you in staying connected to what matters without losing yourself in the process.

This book brings psychological principles from the emerging discipline of climate psychology to provide insights and approaches to help you cope, adapt, and even thrive. The readings and practice exercises will help you reduce emotional overwhelm caused by climate-triggered eco-anxiety, grief, and other expressions of distress while finding ways to participate in the creation of a safer, healthier world. Our planetary health crisis is calling for us to take a deep dive into the true origin of our suffering: our separation from the natural world and the planetary crisis we have created as a result. The readings and practice exercises will help you discover the truth of our profound interdependence with the natural world and to regard it not as something we visit, or to which we connect, but as a living system of which we are an inherent part. You will explore how the cultural delusion of separation from our bodies, from each other, and from the Earth has contributed to environmental degradation and personal distress and how reclaiming our sense of belonging can restore emotional balance and clarity.

By strengthening your felt sense of relationship with the environment, you can cultivate resilience, reduce anxiety, and experience a deeper sense of meaning and connection. Decades of research confirm that even brief experiences of connecting with nature can regulate the nervous system, reduce rumination and stress, and increase prosocial behaviors and environmental stewardship (Berman et al., 2008; Mayer & Frantz, 2004; Ulrich et al., 1991; Weinstein et al., 2009). When we rediscover the visceral memory of being an inseparable expression of the natural world, we support our emotional healing and are inspired to protect the places and communities we love.

CORE THEMES

The emerging field of climate psychology understands that addressing the climate crisis requires both internal and external trauma-informed strategies. Clarifying and then connecting our thoughts, feelings, and nervous system (internal) to social justice and community-oriented engagement in meaningful change (external) are essential paired therapeutic activities. We see frequent examples of how the use of just one of the two strategies (i.e., only internal-informed strategies or only external-informed strategies) creates inherent problems. For example, youth activists, climate scientists, and environmental workers in government and nonprofit organizations report high levels of burnout and chronic stress related to the ongoing nature of the climate crisis (Pihkala, 2020a; Climate Burnout Report, 2023; Hickman et al., 2021). These frontline workers often lead with external strategies, advocating for policy change, community resilience, or environmental protection, yet they may lack the internal resources needed to support the intense climate-triggered distress that accompanies their efforts (Clayton, 2020; Cunsolo & Ellis, 2018).

We also see people experiencing eco-anxiety or eco-grief who seek relief primarily through internal strategies, such as climate-aware therapy, mindfulness practices, or nervous system regulation. However, without opportunities to participate in meaningful, values-aligned climate action, many continue to feel helpless or disempowered in the face of the accelerating crisis (Pihkala, 2020b; Schwartz et al., 2022).

Research consistently shows that both psychological coping and collective engagement are necessary for reducing distress and fostering a sense of agency in a rapidly changing climate (e.g., Lawrance et al., 2022; Schwartz et al., 2022).

This book will help you do both the internal and external work necessary to emotionally cope with the climate change crisis while working to build a more just and sustainable world. By reading these chapters and completing the practice exercises, you will work to heal yourself, your community, your environment, and your world, from the inside out.

Internal Work

Internal landscape work begins with the assumption that the lived experiences of eco-anxiety and eco-grief, and the many expressions of climate-invoked dread, are a natural, empathic response to an unhealthy and perilous situation. I propose a new definition of *emotional resiliency*. The conventional definition asks how quickly we can bounce back to our emotional baseline after a distressing event. But in a rapidly warming world, there is no solid ground to return to. As climate impacts increase, our emotional distress rises in tandem. This book will help you build a new kind of emotional resiliency, one that cultivates the ability to remain present, open minded, and empathic in the face of increasing stress: an enhanced ability to bear witness and be present to more of our changing world and all the feelings that accompany it. A wide array of tools for self-regulation, derived from research on the brain's neuroplasticity, are provided. A robust range of behavioral science approaches and philosophies, including trauma-informed narrative therapies, ecopsychology, evolutionary psychology, mindfulness practices, neuroscience, systems perspectives, and expressive arts therapies, are interwoven throughout.

External Work

We have long known that taking empowered action to change a source of distress reduces anxiety and depression, and that is especially true with respect to the impacts of our warming world. This book provides external strategies to guide you, the reader, to participate in creating a healthier, safer, more just world. The traditional definition of *advocacy* is expanded to show how everyone can make a difference by drawing on their own interests and talents within their existing circle of influence. And because the climate crisis is rooted in insidious and unjust systems, environmental and social justice are both foundational themes of the book.

PRACTICE EXERCISES

This book is considered a workbook for a reason. Each chapter offers several practice exercises that are designed to help you incorporate important concepts into your daily life. The heart of the book lies in these practice exercises, so I invite you to give them a try. I have intentionally kept the chapters brief so you have plenty of room to explore the exercises at your own pace. As you move through the book, you will discover the practices that resonate most deeply for you. Readers who wish to delve further into the research and scholarly foundations of the exercises can consult the References at the end of the volume.

The practices are adapted from the following therapeutic approaches, which are supported by decades of research.

- *Mindfulness* cultivates the awareness necessary for noticing our thoughts and emotions without being swept away by them. Mindfulness has been repeatedly shown to reduce rumination, stress, and emotional reactivity (Creswell, 2017; Kabat-Zinn, 1990).

- *Cognitive behavioral strategies*, an approach supported by decades of empirical study (Beck, 2011; Hofmann et al., 2012), help loosen the grip of patterns that fuel anxiety and hopelessness.
- *Somatic practices*, drawn from research in neurophysiology and trauma-informed regulation, help the nervous system find balance (Porges, 2011; van der Kolk, 2014). Calming our bodies is especially important when climate stress becomes physically activating.
- *Existential exercises* help one address the deeper questions of meaning, responsibility, and belonging that environmental change often brings to the surface (Wong, 2012; Yalom, 1980).

Although these practices originate from different traditions, they share a common purpose: to help us stay present, steady, and connected to what matters most when we are faced with uncertainty or overwhelm. Together, these evidence-based approaches form a comprehensive toolkit for meeting the realities of climate change with clarity and resilience—one grounded breath, one compassionate thought, and one courageous choice at a time.

MY BACKGROUND

I am often asked how I found my way into the practice of climate psychology, which is an emerging specialty in the mental health field. Like many colleagues doing this work, my path is woven from professional, scientific, and deeply personal threads. I have been working as a licensed marriage and family therapist for more than 30 years to help people develop the inner skills they need to face difficult realities with stability and agency. Early in my career, I cofounded one of the nation's first hospital-based integrative medicine programs. My work as a member of an interdisciplinary medical team taught me that true healing happens when we bring multiple lenses to a crisis—not just medical resources but also psychological, relational, and creative ones. That experience laid a foundation that allowed me to bring a psychological perspective into climate conversations where scientists, policymakers, and engineers were already at the table. I saw very clearly that so much of what therapists are trained to do—work with denial, navigate grief and complex emotions, support behavior change, and facilitate difficult relationships—was exactly what the climate space needed.

My own awakening to the severity of climate change began more than 20 years ago, and I shifted both my personal and professional lives toward addressing the crisis. One of my defining moments was when a close friend lost her home in a California wildfire. Witnessing her grief, as well as her resilience, had an enormous impact on me. It led me into training with the Pachamama Alliance (https://pachamama.org/) and into on-site work with a Red Cross Disaster Mental Health team (https://www.redcross.org/about-us/our-work/disaster-relief/disaster-mental-health.html) as well as collaborations with several other groups.

My professional career as a modern dancer dates back to an even earlier stage of life, and I continue to hold a deep belief in creativity as an essential element of our humanity and our healing. Creativity expands what we think is possible, and in the context of climate change it helps us imagine new futures and think beyond existing systems and limitations.

Today, my work sits at the intersection of psychology, climate science, and social change. I teach climate psychology at the California Institute of Integral Studies, consult with environmental organizations and journalists, and speak internationally about how we can stay emotionally grounded while working toward a livable future. I am excited to share with you what we in the field of climate psychology have learned about healing ourselves and healing the world. In the chapters that follow, we will explore together the emotional, relational, and practical tools that make that healing possible.

OVERVIEW OF THIS BOOK

The book begins by tracing the living connection between our inner landscapes—our beliefs, feelings, and nervous systems—and the outer landscapes of a world that is undergoing rapid environmental, social, and political change. In Chapter 1, we are introduced to the full range of climate emotions people may experience as well as the many different ways that climate change is unfolding. We start building a toolkit to help us stay grounded, present, and responsive to both our inner experience and the realities unfolding in the world around us. Uncertainty about how and when climate change will affect us can stir up powerful feelings of dread and disorientation, but it also carries the seeds of creativity. Chapter 2 focuses on how to cope with this uncertainty by inviting us to imagine new possibilities and pathways forward.

Another key aspect of our climate-focused emotions is the ambivalence that can emerge when we try to make sustainable choices while living in a society built on fossil fuels. Chapter 3 helps us navigate through this murky tension and our conflicting feelings without getting stuck or losing our sense of momentum and purpose. In Chapter 4, we turn to *eco-grief*, the deep ache that arises when we face the losses and damage that are transforming our world. The practice exercises in this chapter invoke a tender, embodied presence that supports our heart and body as we explore the various ways that grief makes itself known.

Even when we clearly recognize how the warming world is reshaping our lives, we are all influenced by our *cognitive biases*—those unconscious tendencies that distort, distract, or distance us from fully understanding and engaging with reality. Chapter 5 illuminates these obstructive patterns and helps us uncover them in ourselves and one another with greater insight and compassion. We are aware that a quick technical fix cannot solve the scale and complexity of environmental breakdown; the hard scientific facts have made that abundantly clear. Chapter 6 looks beneath the surface to explore how true change requires a deeper transformation of humanity, a shift in our sense of belonging, imagination, perception, and connection.

Chapters 7, 8, and 9 trace an ever-widening arc of connection. They begin with practical ways to foster community and then move into the deeper belonging we share with the more-than-human world—and, finally, we extend ourselves across time to build intergenerational alliances grounded in eco-justice. Chapter 10 concludes the book by weaving the learning journey into a wider existential frame that touches the deepest roots of what it means to be human. It also offers practical guidance for continuing forward with clarity and intention.

The Appendix contains a curated list of resources: sources of individual and group therapeutic support; tools for parents and educators; and recommended organizations, books, and films that inspire deeper exploration.

COLLECTIVE SOLUTIONS

Although this book provides state-of-the-art tools for transforming climate distress and finding our purpose in this era of climate crisis, it takes the form of an exploration rather than a litany of answers. As human beings, we are on completely new ground when it comes to climate change. As we learn how to work more and more creatively, our collective solutions will gradually reshape the climate crisis and lead us into a future where we can more consciously coauthor our human story.

Whatever has shaped the path that brings you to this moment, it is clear that we need each other if we are to surmount the greatest challenge of our time: addressing the climate emergency. We need everyone's perspective—scientists', Indigenous wisdom-keepers', parents', grandparents', students', activists', economists', medical workers', artists', teachers', policymakers', engineers', farmers', and children's—if we are to understand the entire landscape of the required transformation. All of our talents are essential for creating far-reaching change. Addressing the climate crisis should not rest on the faulty and unsustainable belief that "I have to do it all!" It does ask that we all participate by contributing our time and talents to help build a culture of sustainability within our circles of influence.

Courage. Engagement. Critical thinking. Creativity. Empathy. Follow-through. Self and community care. A bone-deep understanding of our interbeing with all of life. These capacities become our trustworthy guides when we begin working deeply with the many questions and challenges of our era that are brought dramatically into focus by climate change. Let's use every single one of our feelings and experiences to help us grow our resilience and discover how to make sustainable changes for a safer, healthier, and more just world.

CHAPTER 1

OUR INTERNAL AND EXTERNAL LANDSCAPES

Scientists have been warning us for years that climate change is real, dangerous, and relentless, with roots deeply embedded in a wide range of systemic issues driven by human behavior. Current surveys show that 97%–99% of climate scientists attribute recent global warming to human activities (National Aeronautics and Space Administration, 2024). The impacts from our rapidly warming world are part of a *polycrisis* (Cantor, 2025), a term popularized by Adam Tooze, chair of the history department at Columbia University, who was named one of the top global thinkers of the decade by *Foreign Policy* (Columbia University European Institute, 2019). It describes how climate change is interconnected with, and compounded by, crises in multiple sectors: The climate crisis is also a crisis of public health, civil rights, food and housing security, economics, and fiscal justice, and our psyches are paying a steep price for such a universal and deep-rooted existential upheaval.

In this chapter, we explore these intersecting emergencies and their impact on our emotional, physical, collective, and environmental well-being—specifically, how our internal experiences are inextricably linked to our external environments, why distress in the face of climate change is understandable, and why both internal and external action is needed. We will begin exploring our climate emotions and learning to meet them with steadiness and grounding so that the important work of healing ourselves and our environment can begin.

THE INTERNAL LANDSCAPE: HOW EMOTIONS SERVE AN EVOLUTIONARY PURPOSE

Whether you are experiencing the impacts of climate change directly or are learning about them, when you really understand what is happening to our planet and all of its people, animals, plants, ecosystems, and cultures, that realization cannot help but break your heart and awaken a wellspring of emotions, ranging from grief, frustration, love, and anger to a profound longing to protect what remains. Although these feelings are difficult to endure, they are a healthy part of our evolutionary makeup. They are actually our allies, bringing us a message (McLaren, 2023), one that is warning us to transform ourselves and our societies. If no one had empathy for our world to drive us to change, what do you think would be the outcome if the warming climate were allowed to proceed unchecked?

Emotions evolved as useful responses that helped our ancestors survive, connect, and make rapid decisions in complex environments. From an evolutionary standpoint, emotions act as signaling systems that guide behavior toward life-enhancing outcomes: Fear alerts us to danger, anger motivates us to mobilize resources to confront threats or injustice, and joy and love reinforce the social bonding that is essential for cooperative living (Cosmides & Tooby, 2000; Nesse, 2019). Instead of being arbitrary or disruptive, emotions are biologically sourced information streams that can help us interpret our surroundings, evaluate risks and opportunities, and build relationships that sustain us.

Immediate physical threats and survival needs once primarily triggered emotions, but today humans navigate far more complex social and ecological challenges, making *emotional literacy*—the ability to recognize, regulate, and communicate emotions—an essential skill that supports our mental health, improves problem-solving and engagement, and strengthens interpersonal relationships that are vital to building community (Gross, 2015; Salovey & Mayer, 1990). Skillfully working with emotions allows us to pause rather than to react impulsively, to understand the deeper needs or values they are revealing, and to respond in ways that are aligned with our intentions rather than being driven by habit or overwhelm.

In today's world, where new information, threats, and possibilities emerge rapidly, it is no wonder we can feel a surge of conflicting feelings as our systems continually recalibrate. Adding to the challenge is the fact that emotions rarely arise in isolation: They influence one another through interconnected neural and physiological systems, creating blended emotional states (e.g., grief intertwined with love, fear mixed with hope) that provide us with a more nuanced understanding of our circumstances (Barrett, 2017; Larsen & McGraw, 2011). Recognizing the dynamic aspects of our emotions can help us normalize our feelings and support a more compassionate understanding of ourselves as we experience multiple types and layers of climate-triggered distress.

Research shows that cultivating emotional awareness and emotion regulation skills—as you will be doing throughout this book—can reduce stress, improve resilience, and support adaptive coping across many life circumstances (Chambers et al., 2009; Mennin & Fresco, 2014). When we engage with our feelings in this way, emotions stop showing up as problems to manage and instead become sources of wisdom, motivation, connection, and guidance toward the meaningful changes our lives and our world are asking of us.

THE EXTERNAL LANDSCAPE: LOOMING THREATS IN THE ENVIRONMENT

Scientific assessments show that rising temperatures, extreme weather events, and accelerating feedback loops, like permafrost thaw and ecosystem collapse, are pushing Earth's systems toward critical tipping points beyond which recovery becomes unlikely (Intergovernmental Panel on Climate Change [IPCC], 2023). These shifts are creating enormous social and psychological disruptions for individuals and communities: geographical displacement, local and regional conflicts, unprecedented losses, widespread distress, and increased uncertainty (Clayton, 2020; Cunsolo & Ellis, 2018). The convergence of environmental degradation and deepening human vulnerability underscores why climate change is widely recognized as one of the defining existential challenges of our time (IPCC, 2023).

Truly disastrous damage to our world is being purposefully caused by the coal, oil, and natural gas industries (IPCC, 2023). From exploration to their extraction, transportation, and burning for energy, fossil fuels—coal, gas, and oil—have long been the largest contributors to greenhouse gas emissions and climate change (International Energy Agency, 2023a; IPCC, 2023). What is worse is that fossil fuel companies have known for decades about the environmental and climate risks their operations pose, and internal documents reveal that they understood the science of climate change well before it became widely recognized (Franta, 2021; Supran & Oreskes, 2017). They continue to invest enormous sums of money in government lobbying and public relations campaigns to downplay or distort the science behind climate change, delay regulatory action, and protect their profits (Brulle, 2018; Supran & Oreskes, 2021). These industries have been at the forefront of obstructing global initiatives aimed at resolving the climate crisis (Martinez et al., 2023).

It might seem surprising for a book on emotional healing to discuss the oil industry, but climate change is not just a personal struggle: It is also a social and political issue. Seeking relief from painful eco-emotions without addressing the root causes is not sufficient: It's like going to the doctor for relief from foot pain after stepping on a nail and being offered only pain medication and physical therapy with no acknowledgment of the injury or attempt to remove the nail.

What the World's Leading Risk Experts Are Warning Us About

The World Economic Forum (WEF) publishes an annual *Global Risks Report* that evaluates and highlights the major risks the world faces, taking an intersectional approach that examines the connections among environmental, societal, technological, geopolitical, and economic challenges. The report integrates data on issues, such as biodiversity loss, extreme weather events, financial instability, artificial intelligence–related ethics, international conflicts, social inequality, public health crises, and more. Compiled from surveys and expert insights from leaders in business, government, academia, and civil society, the report provides an overview of the key risks and assesses their likelihood and potential impact.

The latest report (WEF, 2025) summarized these findings:

- **Environmental threats are worsening faster than any other major risks over the long term.** Across all 33 global risk areas surveyed, environmental risks worsened the most in the 2- to 10-year time frame, signaling a deepening vulnerability of Earth systems to climate impacts. In other words, the further we look into the future, the more severe our environmental problems become, which confirms that the natural systems on which we depend are becoming increasingly fragile and stressed by climate change.

- **Extreme weather events are the most severe long-term risk.** Extreme weather has been ranked as the No. 1 most severe global risk we will face by 2035, and experts are expressing growing concern over climate-intensified storms, floods, droughts, and heat waves. In the United States, we are seeing this play out first hand; for example, in May 2024 a record number of storms brought tornadoes, hail, and powerful winds that wreaked havoc on communities and destroyed hundreds of homes in Oklahoma, Kansas, and neighboring states (Weather.com Meteorologists, 2024). Deadly hurricanes and flooding have also hit the southern United States (Thiem & Lindsey, 2024), and destructive wildfires have recently devastated large areas in the Los Angeles area (Seydi, 2025).
- **Biodiversity loss and ecosystem collapse are the second-largest risk over the next 10 years.** Nature is breaking down faster than expected. Plants, animals, insects, and entire ecosystems are disappearing or becoming too damaged to function. This creates a domino effect: When nature collapses, it affects everything on which humans rely, including food, clean water, stable weather, and healthy living conditions. Furthermore, major shifts in the Earth's systems, and increasing shortages of natural resources, are close behind, underscoring how deeply our well-being depends on a healthy, stable natural world. The planet has lost an estimated 69% of monitored wildlife populations since 1970, with some of the steepest declines occurring in Latin America and the Caribbean, where average wildlife populations have fallen by approximately 94% (World Wildlife Fund, 2022). These trends are consistent with broader assessments of global biodiversity loss conducted by major international scientific bodies, including the Intergovernmental Science–Policy

Platform on Biodiversity and Ecosystem Services, which in 2019 warned that 1 million species are now at risk of extinction.

- **Pollution has emerged as a major underrecognized global threat.** Many different types of pollution—including dirty oceans, contaminated drinking water, excess nitrogen, and fast-acting climate pollutants—pose greater environmental and health risks than most people realize. These numbers are supported by research from the Lancet Commission on Pollution & Health, which estimated that pollution is responsible for approximately 9 million premature deaths every year, with the greatest burden falling on low- and middle-income countries (Fuller et al., 2022). This confirms that the climate crisis is also a global health crisis because environmental pollution, especially air pollution, remains one of the leading causes of early mortality worldwide.
- **Natural carbon sinks may reach a tipping point.** Earth–system tipping points—those critical thresholds within climate and ecological systems beyond which rapid, irreversible, and self-reinforcing changes are triggered—could accelerate climate breakdown if crossed (WEF, 2025). A key concern involves the destabilization of natural *carbon sinks*, which are systems in the environment (e.g., forests, oceans, and soils) that remove and store more carbon dioxide (CO_2) from the atmosphere than they release (IPCC, 2021; Lindsey, 2025). These sinks are essential for regulating the global climate because they slow the accumulation of heat-trapping gases. However, as rising temperatures, deforestation, and ecosystem degradation place increasing stress on these systems their capacity to absorb carbon weakens. In extreme cases, they may even shift from absorbing CO_2 to emitting it, creating reinforcing feedback loops that accelerate warming. This dynamic reflects the WEF's (2025) broader assessment that environmental systems

are becoming more fragile and that long-term climate risks are intensifying faster than any other major category of global risk.

The Interconnected But Unequal Global Risks of Climate Change

Climate change is already reshaping lives around the world, and the latest global data show just how widespread and inequitable its impacts are. From extreme weather to rising seas, millions of people, especially in low-income and climate-vulnerable regions, are facing increasing threats to their safety, health, and economic stability. The numbers in the following list offer a snapshot of the scale of this challenge, highlighting those who are most at risk and why urgent collective action is necessary in order to protect communities and strengthen resilience:

- **Billions of people are already living in high-risk climate zones.** More than 3.3 billion people live in places that are extremely vulnerable to climate impacts, such as more dangerous storms, rising seas, droughts, and heat waves (IPCC, 2022). This includes many communities in small island nations, sub-Saharan Africa, the Arctic, Central and South America, South Asia, and Southeast Asia, where climate hazards are becoming more frequent and more severe.
- **Climate disasters hit low-income countries the hardest.** Low-income countries lose about 5% of their national income every year because of climate-related disasters, compared with just 0.5% in wealthier countries (UN Office for Disaster Risk Reduction, 2022). These losses strain national budgets and make it harder for communities to rebuild. Many countries in Africa, South Asia, Central America, and the Pacific Islands experience repeated storms, floods, droughts, and crop failures that are pushing people deeper into poverty (World Bank, n.d.).

- **Developing countries have contributed little to the problem, but they face the greatest harm.** Although developing nations have only produced about 8% of historical greenhouse gas emissions, they are on the front lines of climate impacts (United Nations Environment Programme, 2023). This imbalance between those who caused the crisis and those who suffer most from it is at the heart of global climate justice discussions (IPCC, 2022).
- **Climate disasters have displaced more than 200 million people in the past decade.** Over the past 10 years, climate-related hazards, such as floods, fires, storms, and droughts, have forced hundreds of millions of people to leave their homes, often temporarily (Internal Displacement Monitoring Centre, 2024). In 2021, the World Bank estimated that up to 200 million people could be permanently displaced by 2050 as climate impacts worsen, creating one of the largest climate-driven migration trends in history.

So far, we have been looking at global impacts, which is crucial for understanding how deeply we are interconnected with every part of the living biosphere. The events unfolding in the Amazon rainforest and the warming Indian Ocean are interconnected in complex ways. We are deeply connected to each other and the world around us. We cannot divide up the atmosphere, or stop the circulation of air and the flow of water around the planet we share: These elements truly affect us all. But the same social and ethical issues are also unfolding in our own country, states, and local communities.

> *We are deeply connected to each other and the world around us.*

The Challenges of Climate Justice in the United States

Across the United States, long-standing inequities mean that communities of color, low-income neighborhoods, and many Indigenous nations are facing the harshest climate risks despite contributing the least to the problem. These are not just environmental issues; they are deeply human ones.

- **Pollution burdens.** Many Black, Latinx, and low-income communities live near refineries, highways, and toxic sites, exposing them to dangerous air and water pollution. "Cancer Alley" in Louisiana, where industrial facilities surround mostly Black neighborhoods, is one of the most troubling examples (Terrell & St. Julien, 2022).
- **Extreme weather impacts.** When hurricanes, floods, or heat waves strike, vulnerable communities are hit hardest and recover the slowest. Urban "heat islands" make many low-income neighborhoods several degrees hotter, and limited access to cooling increases health risks (Hsu et al., 2021; Palinkas et al., 2022).
- **Energy inequity.** Many people struggle with energy poverty, living in inefficient homes, paying higher utility bills, or lacking access to clean-energy upgrades. Indigenous communities in particular often rely on expensive or polluting energy sources because of chronic underinvestment (Hernández, 2016).
- **Loss of land and culture.** Rising seas, droughts, and storms are forcing some Indigenous communities to relocate, threatening their cultural heritage and sovereignty. The Biloxi-Chitimacha-Choctaw tribe in Louisiana is one of the first U.S. communities that is already undergoing climate-driven displacement (R. Boyd, 2019).
- **Unsafe water.** Water contamination continues to disproportionately harm marginalized communities, as seen in the Flint water crisis, in which a predominantly Black and deeply impoverished community was exposed to lead-tainted water, resulting not only in widespread physical illness but also elevated

rates of behavioral challenges and trauma-related conditions (Butler et al., 2016).

These realities describe families and communities that are navigating fear, loss, and profound uncertainty. Climate justice asks us to look directly at these inequities and invites us to clearly see and elevate the people most affected—and to build a future in which protection, safety, and resilience are shared by all.

HEALING OUR INTERNAL AND EXTERNAL LANDSCAPES

In this book, we will be learning about how to cultivate a healthy internal landscape (our thoughts, feelings, and nervous system) while also working to heal our external landscape (our entire living biosphere) so that our climate distress is reduced and we become cocreators and stewards of a life-sustaining world. Our internal and external landscapes are inextricably connected: Without caring for our inner landscape, we will burn out. And without making change in the world, we are disempowered. However, participating in change is about more than feeling empowered: It is about experiencing ourselves as part of an interdependent web in which our actions matter, our well-being is linked with others, and our efforts contribute to something larger than any one individual. The practice exercises in this book will help you start or continue your journey of personal discovery and healing. You will strengthen your internal and external resources so you can free yourself from burnout and despair.

> *Our internal and external landscapes are inextricably connected: Without caring for our inner landscape, we will burn out. And without making change in the world, we are disempowered.*

Because a range of emotions is a natural response to the devastation caused by our existential crisis, these feelings will continue to be triggered as long as there are damaging and unjust structures in the world. But understanding and working with our emotions is an essential step toward addressing climate change. We cannot "cure" our climate-related anguish while our world is on fire—but if we move together through despair, dread, rage, and grief, we will be better equipped to cocreate a life-supporting future and change the broken systems that cause our painful eco-distress.

I often hear that taking time for self-care and tending to our emotions is a luxury we cannot afford given the urgent need for climate action—every moment counts, and pausing to tend to our inner lives is time taken away from preparing for future changes and working to mitigate the impacts. But the ongoing, accelerating, and often traumatic changes created by the climate crisis can lead to extremely high levels of distress that can easily compromise our ability to take effective measures. Susanne Moser (2020) a research scientist and former employee of the National Center for Atmospheric Research, conducted a survey of frontline climate adaptation professionals, including policymakers, researchers, and urban and agricultural specialists and found that four out of five respondents reported feeling burned out. It is not possible to constantly expend effort over the long haul without finding ways to recharge—especially when you are burdened with the knowledge that we are living on a sharply rising trajectory of climate impacts.

RECOVERY TIME IS NOT OPTIONAL; IT IS ESSENTIAL FOR CREATING CHANGE

Perhaps we can better understand the importance of self-care if we look at the guidelines of others whose work entails striving to meet very high standards: professional athletes. When an athlete is not meeting their performance targets, one of the first things their coach asks them is whether they are taking their recovery time. Recovery is considered an essential part of an athlete's training, not a break from it. Without taking time for muscles to recover, they can end up with a body that can no longer repair itself. Recovery time is also essential for their mind to recharge; this has been shown to result in fewer injuries and a boost in creativity and motivation, ultimately leading to improved performance (Cunningham & Ansorge, 2025). Our transformative climate work, both inside and out, is no less rigorous in its demands, and it requires the same holistic approach.

One of the most powerful underlying factors contributing to climate change is how radically disconnected we have become from the organic patterns and laws of nature. This disconnect is reflected in the work-without-rest approach we are investigating here. Consider this essential rhythm that recurs throughout nature: the ebb and flow of tides, the waxing and waning of the moon, the turning of the seasons, the rising and setting of the sun, and breathing in and breathing out. This rhythm is an ancient and pervasive cadence that is essential to supporting life. In order to engage with climate emotions and climate work, we need to participate and restore, persevere and play, speak out and listen, know ourselves and connect with others, plan and pause to reassess.

Let's begin with an easy practice for aligning ourselves with nature's rhythms through the natural inhalation and exhalation cycle of the breath. When we are anxious or upset, we often take shallow, uneven, or forced breaths that bypass the diaphragm's natural role in moving air efficiently into and out of the lungs. This strained breathing disrupts the body's physiological balance by increasing activation of the sympathetic nervous system and circulating higher levels of stress hormones, such as cortisol and adrenaline, which keep the nervous system on alert and eventually deplete our energy (Jerath et al., 2006; Porges, 2011). Let's start with a breathing exercise that nourishes our inner landscape.

Practice: Balanced Breathing

Balanced breathing does not have to take a lot of time out of your day. Begin with 2 to 5 minutes per day and increase the time and frequency as the exercise becomes easier and more comfortable.

You will practice focusing on taking smooth and steady inhalations and exhalations of the same length. Pick a pace that feels easy but encourages a somewhat slower and fuller breath than usual. Take a break if your breathing begins to feel uncomfortable. Balanced breathing can be practiced sitting, standing, or lying down.

1. Place one hand on your upper chest and the other on your belly so you can feel the movement of your abdomen.
2. Slowly inhale through your nose for four counts, feeling your stomach expanding and pressing into your hand. There should be minimal movement of your other hand.
3. Exhale through your nose for four counts as your belly falls, and very gently tighten your abdominal muscles at the end of the exhalation, keeping your upper hand still. Normal breathing involves a natural break between breaths, so it's fine to have a short pause between the in-breaths and the out-breaths. Sense the timing of the pause that feels best to you.

After 2 to 5 minutes, bring your hands down to your sides, stop the counting, and allow your breath to find its own rhythm. Pause and notice how you feel.

GETTING TO KNOW YOUR CLIMATE EMOTIONS

Now let's get acquainted with the climate emotions that come up the most for you during that practice exercise. We are not aiming to avoid certain feelings, or to search for the "best" or "correct" ones; instead, think of healthy emotions as "feelings that flow." Like other natural phenomena, feelings are transient, arising and receding if we allow them to arrive and then let them go. There is a common fear that if we let ourselves really feel the fear, grief, or anger triggered by climate change we might fall into a pit of despair that we won't be able to escape. However, just the opposite is true: It is when we try to suppress our naturally occurring feelings that we get in trouble. Research shows that when we push down or ignore our emotions, the cost is high: Emotional suppression is linked to increased depression; substance use; and a range of physical health problems, including cardiovascular issues, digestive disorders, and chronic pain (Gross & Levenson, 1997; Pennebaker & Chung, 2011).

> *We are not aiming to avoid certain feelings, or to search for the "best" or "correct" ones; instead, think of healthy emotions as "feelings that flow."*

Trying to suppress your feelings is like holding an inflated ball under water: Think about all the energy and attention required to keep the ball submerged. Your arm will eventually get tired, and the ball will pop up to the surface despite your efforts. We want to guide our feelings to the surface as we gradually and intentionally release their pressure with care and validation. However, a balance must be struck: You don't want to invite your feelings to permanently set up camp any more than you want to avoid them. As you get more familiar with your full spectrum of climate emotions, and listen for and respond to what they are expressing, your ability to be friendly with a broad range of feelings will increase. You will discover the skills you need to navigate the natural fluidity of feeling states, and you will free up energy that can be used constructively in your daily life—as well as to address the current climate impacts on our planet.

Panu Pihkala (2022) a leading scholar of eco-anxiety at the University of Helsinki, partnered with acclaimed climate journalist Anya Kamenetz, who led the Climate Mental Health Network team, working with him to develop the climate emotions wheel shown in Figure 1.1. The wheel is available in more than 30 languages and

FIGURE 1.1. Climate Emotions Wheel

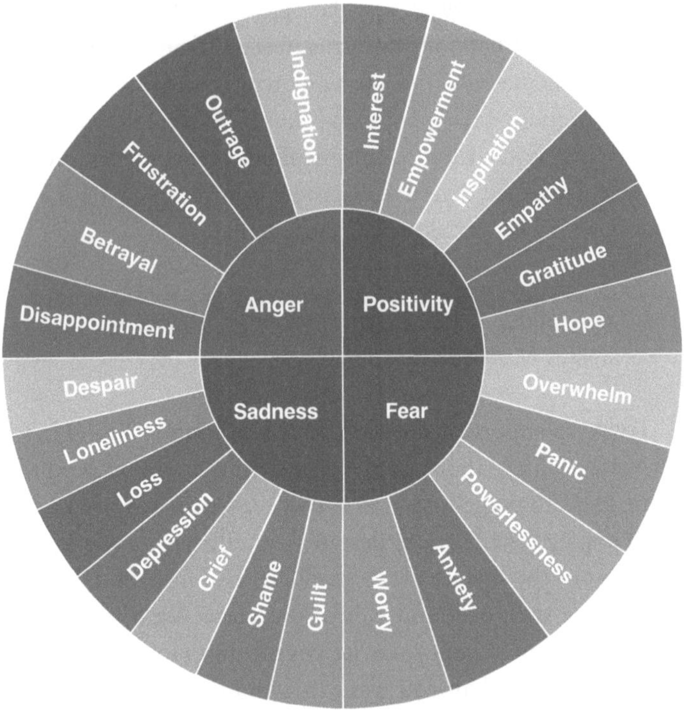

Note. From Climate Mental Health Network. CC BY-NC-SA (https://www.climatementalhealth.net/wheel).

has been used in conferences, schools, libraries, stores, and meetings worldwide. It has even appeared on a video kiosk in New York's Times Square. The climate emotions wheel is a very helpful tool for identifying feelings that are triggered by the climate crisis, the practice, Identifying Your Eco Emotions, is adapted from their work (Pihkala & Kamenetz, 2024).

As you look at the climate emotions wheel and identify the emotions you relate to the most, keep in mind that feelings can shift and change in a day or even a moment. You can also have more than one at the same time. Perhaps you have felt both frustrated at the slow pace of change in critical environmental policies and hopeful about new grassroots efforts, or you have experienced how easily disappointment and anger can pair up, or you have felt rage toward blatant polluters as well as grateful to have found a community of like-minded peers.

Practice: Identifying Your Eco-Emotions

Take a close look at the climate emotions wheel and identify five of the feelings that you relate to the most. If you have an emotion that is not listed on the wheel, include it when you write about your experiences.

1. Are there specific events or experiences that trigger this feeling?

 ____

 ____

 ____

 ____

 ____

2. Does it tend to be accompanied by another feeling at the same time?

 ____

 ____

 ____

 ____

 ____

3. On a scale of 1 to 10, with 10 being the most intense, how strong is it right now?

4. Where do you feel it in your body?

5. If this feeling is a messenger, what might it be telling you?

6. Is there anything that you find helpful when you're feeling this way?

7. If it were available, what's the best possible support you could receive when you experience this feeling? Is there any version of that support you can access right now?

Let's expand our exploration of emotions to include your family and community connections:

1. Does anyone in your family, workplace, or usual social settings ever express climate emotions?

2. Do you sense that anyone in those settings may have unexpressed or concealed climate emotions?

3. Do you have feelings that arise from an empathic connection to others or to the other-than-human world, including geographic locations, cultures, animals, or plants?

____

____

____

____

____

We will be working with all of these feelings, and others, throughout this book, but you have already completed an important step by becoming more familiar with your own inner landscape. You may feel relief just acknowledging these feelings and knowing that many others experience them too. Or you might find that your feelings have intensified because you have been focusing your attention on them and the triggering circumstances. As important as it is to open up to these feelings, it also is essential to know how to ground and contain them. There are two parts to the next practice exercise: Give them a try and notice any shifts in your experience.

Feel free to modify the steps in the following exercise if your circumstances call for a change in the instructions. For example, if you are deaf or hard of hearing, skip Step 3 in the grounding part. Or, in the closing-breath step at the end, if it makes more sense to breathe down your spine or into your hands rather than into your feet, know that the practice will be just as effective with these adjustments.

Practice: Containing and Grounding Your Emotions

Containment

Choose a box, jar, safe, or any other handy container that you can open and close. Write a word, phrase, or brief description of any feelings you are still experiencing after the previous eco-emotion exercise on small slips of paper. The notes can include descriptions of sensations in your body and active thoughts. You can also make simple drawings that express your feelings. As you place the completed notes into the container, imagine that you are storing the feelings along with the notes, knowing that they will be held there for safekeeping until you choose to revisit them. Close the container, or put on its lid. Find a safe place to store your contained feelings—for example, on a shelf, in a cupboard, or in a drawer—and feel a sense of completion as you walk away for now.

Alternatively, rather than using a physical container, just imagine one. It can be any size, color, texture, or shape you wish. How thick are the sides, and what are the texture and design like? Image transferring your thoughts, feelings, and sensations into your container for safekeeping, and find a place where it can rest until you want to revisit those experiences.

Grounding

This 5–4–3–2–1 grounding practice activates your senses and guides your attention from your internal landscape to connect it with your external surroundings and the present moment. It can be practiced at any time and anywhere. As you go through each step, acknowledge each item by silently naming it in your mind or by saying it aloud. Try to avoid adding a mental storyline or extra details; instead, keep it simple.

1. Acknowledge five things you can see.
2. Acknowledge four things you can touch.
3. Acknowledge three things you can hear.
4. Acknowledge two things you can smell.
5. Acknowledge one thing you can taste.

Close this exercise by taking three full breaths while imagining that your exhalation is traveling down your body, out through your feet, and entering the Earth. Feel yourself being present in your body as you move on into the rest of your day or evening.

JOY

There is one emotion not included on the climate emotions wheel that is important to explore, and it is not one that usually comes to mind when we think of climate change: joy. It can be hard to imagine joy associated with the existential dread evoked by the climate crisis. We often talk about our efforts to making the world healthier and more just—but Gerald Torres (2023), Professor of Environmental Justice at the Yale School of the Environment and the Yale Law School, took it even further: "The future we want is in our hands, but it's not going to be done if we accept defeat or think we can't make the change. We have the capacity to produce a joyous world" (para. 3).

Most of the time our eco-thoughts and feelings are bound to the painful realities unfolding on the planet, and we can work to develop skillful ways to more consciously shape our relationship to those events. But what if joy is free of specific inner or outer conditions and instead is a part of our very essence, our life force—our essential nature? What if we can move into joy from the ever-available experience of awe? For example, in those moments when we gaze up at the night sky and take in the expanse of the universe, we might realize how rare and unique we are and yet so small. What would it be like if we brought joyful awe to our relationships and to our work for the planet? Perhaps poet, professor, and equal-rights activist Audre Lorde had this in mind when she described joy as the energy for change (Lorde, 1984).

Lorde (1984) also described joy as an "act of resistance" (p. 53) because it supports our unity and counters the control of the oppressive structures that divide us. Her ideas are related to an emerging orientation, advanced by adrienne maree brown (2019), called *pleasure activism*. brown challenged us to rethink the ground rules of activism and awaken our desire to make social justice work the most pleasurable of human experiences as we choose to settle for nothing less than a fulfilling life.

Practice: Welcoming Joy

Pause for a moment and imagine that your breath could move in and out directly through the center of your chest, near your heart. Close your eyes, and after about 2 minutes of this "heart-breathing," call to mind a person, place, or circumstance that naturally elicits a feeling of joy, appreciation, gratitude, or love. Keep imagining the person or situation for another 2 minutes, and notice the feelings it generates in your body, mind, and emotions.

Now release your focus on the specific image, but see if you can maintain your attention on the feelings.

Next, bring the pleasant feelings you have just discovered into your interactions with others and whatever form of climate engagement you anticipate doing. Play it out in your mind's eye like a dress rehearsal. When you are finished, open your eyes.

Rather than guiding you toward a single feeling, this practice exercise invites you to tap into the deep well of your humanity and develop your ability to connect with and become attuned to others. If this feels like a practice that could be beneficial, add a note on your calendar where you have scheduled your next climate action to remind yourself to take a moment before the gathering to revisit this practice.

Transform your experience into a writing practice, filling in the blanks with what you feel or can imagine. Feel free to add additional lines to the following poem:

When the world feels cold and dark,
Joy is here.
Through the silence, through the fight,
______________________ is here.
When ______________________,
Love is here.
In every scar, in every star,
______________________ is here.

As we continue to explore the climate crisis and all the thoughts and feelings it evokes, we will look more deeply at how we can embrace them instead of lashing out or collapsing. Caroline Hickman, Professor of Climate Psychology at the University of Bath in the United Kingdom, summarized it this way: "We need to not only *grow up* in the climate crisis by cultivating our imaginative, creative, determined, and hopeful capacities, we also need to *grow down* by building our tolerance for guilt, shame, anxiety, and depression" (quoted in Wray, 2020, para. 10). In Chapter 2, you will discover strategies for managing uncertainty, a major aspect of climate change that intensifies our sense of distress.

CHAPTER 2

NAVIGATING UNCERTAINTY

No matter what aspect of climate change or its impacts we encounter, a psychological factor that often intensifies our emotions is the discomfort triggered by uncertainty. By learning to navigate this underlying stress amplifier, we can reduce the distress we feel about climate concerns and become more capable of participating constructively in a changing world. Let's first explore how to best face the unknown and our frightening climate-changed future.

We humans are wired to fear the unknown. That innate fear is designed by evolution to do an important job. It is a complex biological response that activates multiple regions of the brain, including the amygdala in the limbic system, as well as areas of the neocortex, which trigger a surge of chemicals that prepare us to fight, flee, freeze, or appease in the face of perceived threat (LeDoux, 1996). It may be common sense that tells us to avoid a shortcut through a dark alley or sidestep a big spider, but an ancient voice, deep in the recesses of our brains, also is weighing in.

Climate change keeps many of us on edge, constantly wondering what might happen next. As we learn more about the predicted threats—rising temperatures, fiercer storms, increasing wildfires, and disappearing ecosystems—it is natural for our bodies and minds to stay on high alert. This ongoing sense of uncertainty can lead to a chronic state of *eco-anxiety*, which, as mentioned in the Introduction, is a term used by researchers to describe the worry, fear, and grief associated with environmental change. Although eco-anxiety is not an official clinical diagnosis, it is a real and increasingly common emotional response to the climate crisis. It reflects our deep instincts to seek safety, protect what we love, and make sense of a rapidly changing world (Cunsolo & Ellis, 2018; Pihkala, 2020a). Although we have acknowledged that these fears are completely normal, they can escalate and eventually overwhelm us unless we learn how to work with them.

The multitude of ways that the impacts of climate change may unfold are fraught with uncertainty. The future of our planet depends largely on how much greenhouse gas is emitted into the atmosphere, so whether, and when, carbon reduction or elimination policies and practices are put into place will be critical. Climate change operates within complex, interconnected systems that include the atmosphere, oceans, ice sheets, ecosystems, and human societies, all of which influence one another in nonlinear ways. Small changes can therefore trigger disproportionately large effects, in particular once critical thresholds are crossed. For example, warming in the Arctic has accelerated the loss of sea ice, which in turn reduces the Earth's ability to reflect sunlight, amplifying further warming—a feedback loop that has progressed faster than many models initially projected (Lenton et al., 2019). As a result, even climate scientists are often surprised by the pace and scale of change, reflecting the inherent difficulty of modeling complex systems with so many interacting variables and potential tipping points (Lenton et al., 2019).

In this book, we will learn how to work with the fears that climate uncertainty can easily generate. This is far from a "keep calm and carry on" approach, which can imply a toxic positivity that encourages business as usual; instead, this is about increasing our capacities in our personal, professional, and civic lives and growing the skills we need for creative thinking, building healthy relationships, learning useful communication, and taking targeted action. It is helpful to look at this journey through a trauma-informed lens known as the *window of tolerance*, a concept popularized by psychiatrist Dan Siegel (1999, 2012), who expanded on earlier research about how our minds and bodies manage stress and emotional arousal.

THE WINDOW OF TOLERANCE

When we are within our window of tolerance, our sympathetic (activating and alert) and parasympathetic (calm and restful) nervous systems are in balance. These systems are continually microadjusting themselves, and each other, and this synergy helps us respond to the inevitable ebb and flow of the situations and feelings we experience throughout our daily lives. We can manage and recover from challenges and stressors without much difficulty. There is a sense of ease and competence that develops as we learn, play, work, love, and rest.

However, when there are too many sources of stress, or a major disruption in our lives, we can be pushed into a state of either *hyperarousal*—reactive, panicked, or impulsive—or *hypoarousal*, when we become numb, checked out, or isolated (Schauer & Elbert, 2010). Our brains have a threshold for the amount of distress we can tolerate, and once we are past that threshold the prefrontal cortex, which helps us regulate emotions and distress, begins to shut down (Arnsten, 2009). When this happens, there is often damage done: We might lash out in a relationship, seek addictive substances to numb out, or disappear from a situation that needs our attention, all of which may escalate the problem. Thus, the window of tolerance is the zone between hyperarousal and hypoarousal in which we can best meet challenges with awareness and flexibility (Siegel, 1999; see Figure 2.1). Many of the practices you will be learning throughout this book will help you lower your level of distress and return to your window of tolerance.

The size of your window of tolerance is not fixed: It can shrink or expand on the basis of a variety of factors. We are born with different characteristics that influence our window of tolerance, and numerous life experiences can narrow or broaden it. For example, personal and generational trauma, unresolved grief, a lack of social support, or even poor sleep habits can all act to shrink our window. When your heart and mind are already at maximum capacity, there is less room for facing additional concerns without attending to the existing stressors first. However, as you learn to recognize, expand, and regulate your range of tolerance you will have a much greater

FIGURE 2.1. The Window of Tolerance

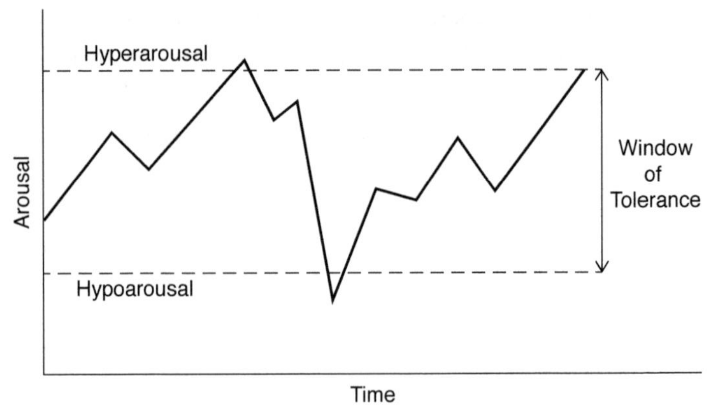

capacity to manage your emotions and respond in ways that are life enhancing for yourself and those around you.

Emotional resilience has traditionally been measured by how quickly someone can bounce back to their usual routines and emotional state after a stressful event (Smith et al., 2008). However, our rapidly warming world and the challenges posed by climate change push us to rethink this. Given our current trajectory, it is clear that we will face increasingly significant challenges and that there is no returning to the old normal. Resilience now means building our ability to embrace more: more change, more uncertainty, more distress and, hopefully, more joy. Growing our resilience today involves expanding our window of tolerance, a process that is accessible to all of us, regardless of our past experiences, temperament, or current circumstances.

> *Resilience now means building our ability to embrace more: more change, more uncertainty, more distress and, hopefully, more joy.*

EXPLORING YOUR WINDOW OF TOLERANCE

Let's begin by exploring our own windows of tolerance and familiarizing ourselves with practices that can help strengthen our resilience. We all have moments in our lives when we were less than our ideal selves, and it is not easy to revisit times that summon regret or even shame. But there is tremendous value in reviewing these experiences through a lens of gentle curiosity and learning. The more we can understand ourselves—our habits, vulnerabilities, beliefs, tendencies, wounds, and strengths—the more we can come home to ourselves and be in healthier relationships with others, build community, and live in reciprocity with the living Earth. So, as you complete the following exercise, try to remain open and curious.

Practice: Exploring Your Window of Tolerance

We can all move into a state of both hyper- or hypoarousal, but a person usually has a tendency toward one or the other. What is your most common direction in times of high stress? Reflect for a moment on your past experiences and estimate the percentage of time you default to hyperarousal (lashing out) or hypoarousal (checking out).

Hyper: ___%   Hypo: ___%

Now write about a recent time or a significant experience, perhaps climate triggered, when you moved outside your window of tolerance. It is beneficial to do this exercise more than once, so begin gently. Instead of working with the most intense memory you can recall, start with a less upsetting event.

1. What happened?

____

____

____

____

____

____

2. What triggered it?

____

____

____

____

____

____

3. How did you feel at the time?

4. What thoughts were running through your mind?

5. What did you do?

6. What were the consequences?

7. How did you feel after the intensity passed?

8. If you could relive that experience in an ideal world, how would you want to act if you knew what was possible? The situation may still be triggering, but what are other potential ways to respond?

EXPANDING YOUR WINDOW OF TOLERANCE

Beginning to see healthier alternatives for dealing with a high-stress situation primes our neurological pathways for resiliency as we navigate our changing world. Imagine rating your level of emotional distress on a scale that ranges from 0 to 10, with 0 representing smooth sailing and 10 representing emotional turbulence and a buildup of distress that pushes you to the edge of your ability to cope. Think of this scale as a navigation tool—a way of staying oriented when the emotional terrain gets rough. If we can recognize when we are escalating from 6 to 7, or from 7 to 8, and we begin applying one of the tools provided in this chapter, it will be much easier to stay within, or expand, our window of tolerance. Once we have hit 10 or beyond, it is much more challenging—if not impossible—to work with our inner experiences in a skillful way until after the intensity fades. Imagine trying to stop a train that is barreling at full speed down the tracks: The effort to slow down takes much more time, space, and effort because of the high speed and momentum. Our emotional velocity is much the same. Developing awareness of our distress level before it peaks is itself a form of practice, one that strengthens our ability to remain present and responsive.

To expand our window of tolerance, we can learn to identify the signs and signals that we notice when we are moving beyond a 6 on our distress scale. These will be different for each person, so reflect on your own experiences. Some people notice physical cues first: a tightening in the chest, clenched jaw, or shallow breathing. Others pick up on shifts in thought patterns, like catastrophizing or a sudden urge to disengage. The next practice exercise provides a few examples.

Practice: Recognizing Opportunities for Change

When your distress rises above a 6, what behaviors or feelings do you have? In the lists below, circle the ones that apply, and fill in the blanks with any other personal signs.

Hyperarousal (lashing out)	Hypoarousal (checking out)
Teeth-grinding	Daydreaming/disconnected
Elevated heart rate, tension	Eating salty or sugary snacks
Pacing/restlessness	Losing track of a conversation
Interrupting others	Feeling zoned out, spacey
Shoulder tension	Internet scrolling
Irritability/impatience	Distracted
Impulsive/defensive	Withdrawn, shut down
Shouting	Difficulty making simple decisions

Throughout this book, you will be consistently adding to your resiliency toolbox, but let's start with the resources you have used to get through challenges before because building resilience often starts with recognizing what is already within reach. For hyperarousal (lashing out), you will want activities that are soothing or grounding because the sympathetic nervous system is overstimulated. This can include taking a walk, doing some deep breathing, and listening to relaxing music. Sometimes it is helpful to burn off excess energy with something more vigorous, like kickboxing or running. For hypoarousal (checking out), activities that bring you into the present moment and are gently invigorating are helpful, for example, dancing or exercise, a meaningful conversation with a friend, cooking.

Practice: Bringing You Into Your Zone of Tolerance

Fill in the chart with familiar approaches that appeal the most to you, such as reaching out to a friend, doing a brief breathing practice, taking a walk outside, journaling for a few minutes, or any other strategies that help you feel grounded and supported. Sometimes we already have helpful tools but, in the moment we don't think about using them. Writing them here can serve as a useful reminder, especially when you are in a state of distress.

Strategies for hyperarousal	**Strategies for hypoarousal**
______________________	______________________
______________________	______________________
______________________	______________________
______________________	______________________
______________________	______________________
______________________	______________________
______________________	______________________
______________________	______________________

ACCEPTING UNCERTAINTY

Kaira Jewel Lingo is a Buddhist teacher in the Zen tradition of Thích Nhất Hạnh and the Vipassana insight lineage. Her exposure to social justice started early: Kaira was born into an interracial family in Chicago that focused on service to the poor and marginalized. In her work, she explores the intersection of racial, climate, and social injustice, and she feels especially called to share the Dharma with Black, Indigenous, and other people of color as well as activists, educators, youth, artists, and families (Lingo, n.d., 2021).

In her book *We Were Made for These Times: Ten Lessons for Moving Through Change, Loss, and Disruption*, Kaira wrote the following:

> It's clear to me now that there is no other time I should have been born into. While the basic freedoms many enjoy today are due to the enormous sacrifices of the many young people of the Civil Rights Movement, I believe we are each called to continue their work in new ways to bring about even deeper freedom and greater awakening so that we can truly live harmoniously and altruistically with each other, our world, and all its species. (Lingo, 2021, p. 110)

In an article for *Lion's Roar* magazine, she added,

> We must respond. But how? And where does that response come from? The way we respond is crucial. . . . In every step we have to manifest what we want, rather than run toward something we hope will happen in the future. The future is made up of this moment. . . . We can deeply attend to our breath and our steps, knowing this will only strengthen us for the significant work ahead. (Lingo, 2020, paras. 7, 10, 30)

Kaira invites all of us to learn to move through intense and challenging times with mindfulness. With respect to living with uncertainty, she described the value of accepting the space of "not-knowing," which offers boundless potential for creativity and insight, suggesting that rather than fearing uncertainty, we "embrace it and all its gifts" (Lingo, 2021, p. 21). Building on this idea, we might draw inspiration from the London Fire Brigade's ethos, captured in the phrase "Love is the running towards" (DNCO, 2022), which invites us not only to accept uncertainty but also to meet it with courage and an open heart.

We don't usually realize it, but we are all very experienced at living with the unknown. We begin each day not knowing what will happen beyond the moment we are in right now, starting when we first wake up. And then there's the next moment. And then the next. And then the next. They may feel like they are running together in a blur of consistency, and our brains assume that "next" will be much like "now": Our lives are often scheduled and planned, so that is frequently the case. Pause and consider how "what's going to happen next" is actually hidden behind a veil of assumption. The veil is meant to be helpful, a way of filtering out the vast possibilities so that day-to-day life is more manageable. But it is also somewhat stupefying: We are lulled into a half-conscious view of reality that mutes both the frightening and awe-inspiring features of life. That is why an unexpected event can feel so shattering: The veil of assumption is suddenly ripped away, and we feel exposed, vulnerable, and disoriented. Suddenly, the life we have been living is forever changed, and everything feels very different. These moments can shift the arc of our lives: We meet our life partner, receive a medical diagnosis, launch our career—or discover that we are living in the midst of a climate crisis. Although the brain's veiling mechanism and our eager embrace of it serves us in many ways, it is devastating when it blinds us to the truth of climate change, which requires an eyes-wide-open awareness.

We don't usually realize it, but we are all very experienced at living with the unknown.

There is an art to being present in the moment and navigating change, and we can develop more agency with our feelings rather than simply remaining at the mercy of our brain chemistry. We can make conscious choices to shape our lives: In fact, in this time of crisis, we must.

Life is more like the flow of a river than the constant state of solid ground. When we shift our orientation to the dynamic nature of the living world of which we are a part, we can live much more effectively. Consider novice sailors who drift with the movement of the tides when they are first learning. They know there is no controlling the weather, but they gradually learn to recognize the patterns of the sea currents and the wind. They acquire the proper equipment and discover how to operate the wheel and rudder. They gradually learn to maneuver the boat with greater precision and nuance, leveraging the wind's energy and becoming more and more responsive to its ever-shifting changes. Although they cannot see the destination for 99% of the trip, they arrive by referencing a compass that points true North.

If we want to develop the confidence and competence to take the helm of our ship and face the uncertainties and unknown forces throughout our climate change journey, we need to become aware of the shifting swells and currents of our dilemma and develop a similar attentiveness. We need to find the true North of our priorities and values. We can learn when to change our tack, when to batten down the hatches, and when to ride things out.

NOTICING THE CURRENTS

Have you ever eaten a snack and didn't register the tastes and textures because you were mentally revisiting a recent conversation? Have you driven to your destination but were not sure how you arrived because your mind was busy planning an upcoming event? You are not alone:

We all have thoughts pulling us into past- and future-thinking with little to no awareness of our bodies or surroundings. However, being caught in the constant current of distraction and mental noise can activate the body's stress response, raising levels of stress hormones, like adrenaline and cortisol. When these levels stay elevated for too long, they contribute to inflammation, higher blood pressure, and chronic distress (McEwen, 1998). We also miss out on important or precious moments in life because there are so many things we can see and feel only when we slow down and are present for them. Although the primary purpose of mindfulness is to build the agility to respond effectively to emerging situations by being more connected to our surroundings, it is also a resiliency tool. Research reveals that present-moment mindfulness boosts mental health, improves focus and memory, increases the quality of relationship interactions, and even improves sleep (Creswell, 2017).

One of the most important ways to attune to the living world and increase our awareness of the currents of change is by heightening our sensory awareness. Although doing this is a simple practice, it is not always easy at first. If you are new to mindfulness practice, you might want to start with 5 minutes and eventually add more time to your practice. Spend about 1 minute on each of the steps in the following exercise.

Practice: Entering the Present Moment Through Sensory Awareness

First, find a comfortable place to sit. This is a practice that you'll ideally do in a safe and beautiful outdoor location, although it's possible to do it anywhere. At the heart of the exercise is paying attention to your physical senses and noticing when your attention gets pulled away from your senses and into your thoughts. Whenever that occurs, simply redirect your focus to your sensory experience. This will likely happen many times, even during a brief practice session. The more often you notice and return to your senses, the more you'll build a muscle, so to speak, that helps maintain present-moment awareness.

1. Sensation/texture: Notice the physical sensation of all the areas of contact between your body and the things supporting it: the wood or cool metal of a bench, a rock, the tree bark behind your back. Note the types of sensations: light pressure, warmth, heaviness, or lightness. And textures: Do they feel smooth, rough, scratchy, or solid? What do you notice about the air on your skin?
2. Sound: Focus on the sounds you hear, both close up and at a distance: birds, muted conversation, machines. And nearby: rustling insects, a breeze in the trees, a stream, the crunch of leaves of someone walking, your breath.
3. Scent: Do you smell soil, flowers, or simply a kind of freshness? Or perhaps the aroma of cooking nearby.
4. Visual: Really look at your surroundings. What colors do you notice? Are there multiple shades of greens, browns, yellows, or blues? Note the quality and shifting patterns of light.
5. Movement and connection: Now take a moment to be with it all, like a symphony of the senses. Notice how there is a constant play and movement from one sense to another: the dissolving of a cloud, a falling leaf from a brief gust of wind, the chittering of a squirrel—all the currents of now.

Reflect for a moment after your practice. How do you feel? Did you notice anything new?

____

____

____

____

____

In the practice of being in and with the natural world, perhaps you noticed a sense of connection or belonging that went beyond simply witnessing and enjoying natural beauty. Belonging is a relationship, one that requires reciprocity to thrive. The threads of connection don't stop with what you can see, hear, and smell. The very air you breathed during this last practice exercise mingled with the shrubs and trees and traveled beyond what you could see or hear. And if you pause for a moment, you might even wonder where your breath has traveled and what beings and landscapes it touched on its way to you and on its way out again. Where is it now? Air doesn't stay still for long. Researchers estimate that wind currents can move and mix air around the globe in just a few weeks, roughly about a month, on average (Prather, 2025). Later in this book (Chapters 7–9) we will explore how the different qualities of belonging nurture our collective belonging, which we also need to sustain our lives.

Throughout this book, you will discover more strategies for dealing with uncertainty by focusing on what you can control; connecting with others who are also dedicated to positive change and making an impact; and breaking things down into smaller, manageable steps. Take one step at a time, and celebrate your progress along the way.

In Chapter 3, you will learn ways to identify, express, and work with ambivalence, which will make it much easier to move through it in a conscious and nonshaming way while you work toward change. Although this takes skill and practice, learning to work with ambivalence is an essential part of navigating our climate emotions—and a natural progression from living with uncertainty.

CHAPTER 3

RESOLVING AMBIVALENCE

As caring people who want to serve as responsible Earth stewards, it is common to experience a disconcerting conflict between our lifestyle choices and their impacts on the environment, and we are often at a loss for how to resolve this conflict. For some of us, this can be compounded by the fear that changing our lifestyle will mean giving up hard-earned comforts and that ultimately it might not make a significant difference in healing our world. And it is often the case that eco-friendly options simply are not available. These feelings of *ambivalence*—wanting to make a difference but feeling overwhelmed or hopeless—make it very difficult to work for social change. In this chapter, we explore why ambivalence is so detrimental and how we can resolve it to move forward.

AMANDA'S STORY OF AMBIVALENCE

Amanda came into the therapy room and flopped into a chair looking defeated.[1] I could see shifting emotions of sadness, defiance, and exhaustion in her eyes as she announced, "The blueberries did

[1]Amanda is based on a real client, but her name and some identifying characteristics have been changed to protect confidentiality.

me in!" Amanda was a university student in her late 20s who was studying environmental science. She had initially sought therapy to help with the eco-anxiety triggered by what she was learning in her degree program. After working to develop self-regulation skills and reaffirming her commitment to a career path focused on restoring ecosystems, our most recent sessions had been devoted to her interest in building a healthier and more sustainable lifestyle. Amanda was well informed about the complexities of our global carbon impacts and their many environmental consequences, but she had not yet encountered any big challenges in the microcosm of her daily life. Now I was curious to hear about the blueberry issue!

Amanda had been carefully tracking her fossil fuel consumption and was determined to make realistic but meaningful changes. She recently had allocated a larger portion of her budget for buying organic produce whenever she could, which was not easy for a student. Her "blueberry defeat" had occurred at a local grocery store: She had read the label on a container of organic blueberries and discovered that they had been flown in from South America and were packaged in nonrecyclable plastic. After weeks of grappling with the inescapable fact that fossil fuel consumption is unavoidable in the course of meeting our basic needs, this felt like a breaking point. Overcome with frustration, she had put the blueberries back on the shelf and left the store in tears.

Climate ambivalence arises when we struggle with the tension between our desire to make environmentally conscious decisions and the practical or systemic obstacles that make those choices challenging, inconvenient, or even ineffective. For Amanda, the inability to align her daily activities with her dedication to healing the Earth resulted in intense frustration, eroded her energy, and briefly undermined her resolve. Sadly, these kinds of dilemmas are both insidious and widespread. Here are just a few of many examples:

- **You want to reduce energy consumption, but you rely on energy-intensive technology.** You might wish to reduce your carbon output, but at the same time you rely on items with energy-intensive consumption, such as computers, artificial intelligence, and air conditioning units, especially if your home's heating or cooling system runs on fossil fuels.
- **You aspire to a zero-waste lifestyle, but you struggle with packaging.** You want to reduce waste and stop using plastic, but you struggle because almost all the products you purchase, from groceries to toiletries, are packaged in plastic or other unsustainable material, and you have limited access to bulk-buy stores.
- **You'd like to eat a plant-based diet, but local options are lacking.** You'd like to adopt a plant-based diet, but it's difficult because of a lack of local, affordable, or fresh plant-based choices where you live. Some plant-based foods are also highly processed, and many are transported over long distances.
- **You prefer a walkable lifestyle, but you live in a car-centric city.** You'd rather walk and bike more and drive less, but you're forced to drive because you live in a car-centric area with limited sidewalks, bike lanes, or public transport.
- **You drive an electric vehicle (EV), but you charge it from a nonrenewable grid.** Perhaps you could own an EV to reduce emissions, but your electricity is primarily sourced from plants that burn coal or other fossil fuels, which undermines some of the environmental benefits of your EV.

WHEN AMBIVALENCE BECOMES PARALYZING AND COUNTERPRODUCTIVE

Climate psychologist Renée Lertzman is widely recognized for highlighting the pivotal role ambivalence plays in how we engage with climate change. She explains that ambivalence, our inner conflict between caring and feeling paralyzed, emerges as we navigate the realities of a fossil fuel–dependent society while striving for transformation. According to Lertzman (2023), it is only by acknowledging and working through this ambivalence that we can move beyond the draining cycles of guilt and shame and begin to access meaningful solutions. Much like overwhelm, she notes, the weight of unresolved ambivalence can be mistaken for apathy, especially when we feel stuck or frustrated. As we saw with Amanda, ambivalence can become paralyzing and counterproductive. It is easy to feel like giving up when our efforts seem pointless, but remember: The corporations that want to continue using fossil fuels as our primary form of energy generation count on exactly this reaction. It is part of a deliberate strategy to keep us overwhelmed, confused, and demoralized—but we are learning to be wise, determined, and resilient.

We have been exploring the benefits of naming and acknowledging our eco-emotions as a way of processing and managing them. Let's apply this to identifying your own areas of climate ambivalence. Perhaps you identify with one of the examples listed earlier. Maybe it shows up in your relationships, when you feel torn between loyalty to a friend and your belief that they are doing considerable environmental harm. Or you are a member of an environmental advocacy group but have doubts about their methods or leadership. You may feel compelled to join a civic demonstration, but at the last minute you stay home out of fear of potential dangers. You would like to speak out at work about wasteful practices, but you wonder if that will put your reputation, or even your job, at risk. Perhaps your partner doesn't share your level of care and commitment to sustainability, so you experience both love and frustration, commitment, and resentment—and everyday choices are turning into an exhausting source of tension in the relationship.

Practice: Identifying Areas of Climate Ambivalence

Reflect on the following questions, and write about your own experience.

1. What are some of the elements of a sustainable lifestyle that you aspire to adopt?

2. Do you feel torn about any of those activities or commitments?

3. Identify each side of the tug-of-war for one of the things you identified.

4. Are there any areas of your life where you don't feel fully committed to making sustainable changes?

____

____

____

____

____

____

5. Has your ambivalence caused you to avoid or procrastinate about certain things?

____

____

____

____

____

____

6. Are there areas of your life where you don't feel like you're being true to yourself? What do you think you should want?

____

____

____

____

____

____

FLEXIBLE VERSUS DICHOTOMOUS THINKING

Ambivalence is a common emotional experience, especially when we are making decisions or judgments in new or unknown situations, such as those related to climate change and sustainability. Although engaging with our ambivalence can be uncomfortable, it can lead to more thoughtful and intentional choices. It encourages us to pause and reflect about competing options and conflicting feelings. This is especially important in complex social situations, for example, when loyalties are divided in groups or relationships. Exploring our emotions and values more deeply can help bring clarity, offer us the chance to reconcile opposing needs, and ultimately lead to rewarding ways to move forward.

One of the key elements in working through ambivalence is *flexible thinking*: the ability to adjust our thinking in response to new information, changing circumstances, or different perspectives. It helps us consider alternative solutions and modify our mindset when necessary. Responding to the complexities of climate issues requires inventive problem-solving. Flexible thinking lets us ask "What is the best option available in this moment?"

In contrast to flexible thinking, *dichotomous thinking* (black-and-white thinking) views situations, people, or outcomes in extremes—either all good or all bad—with no room for middle ground or nuance. This mindset leaves little room for recognizing partial success or progress. For example, if a goal is not met perfectly, you might see it as a complete failure even if the result has value. In the following exercise you can work on some of the dilemmas you identified in Practice: Identifying Areas of Climate Ambivalence.

Practice: Considering Opposites With "Even-As" Thinking

This is called the *"even as" practice*, and it is a reflective exercise that encourages us to embrace multiple and often-contradictory truths simultaneously. It helps us view situations in a nonbinary and expansive way and recognize that it is natural to live with a mix of conflicting or complex thoughts, emotions, and situations.

Reflect on the areas of ambivalence you wrote about in Practice: Identifying Areas of Climate Ambivalence, and provide examples in which there are multiple truths at play. Start by addressing the more troubling aspect, then follow it with the lighter or more beneficial part that also holds true. Here are two examples:

- *I experience intense periods of eco-anxiety* even as *I am grateful for moments of calm in my life.*
- *I am fearful of the future* even as *I continue to work for change.*

____________________ even as ____________________.

____________________ even as ____________________.

____________________ even as ____________________.

____________________ even as ____________________.

____________________ even as ____________________.

____________________ even as ____________________.

This flexible mindset is especially relevant when we consider climate mitigation and adaptation. Numerous experts, including renowned climate scientist Michael Mann (2023), have used the metaphor of climate solution off-ramps that can be used "even as" we doggedly pursue the long game of reversing climate change rather than embracing the all-or-nothing view of climate change driving us off a cliff. Although we may not always be able to achieve our most ideal vision, making incremental improvements that reduce harm is still valuable and helps us preserve what we can while we continue to work toward more ambitious goals. Progress counts, even when perfection is not possible—and perfection is never possible.

> *Progress counts, even when perfection is not possible—and perfection is never possible.*

HARM REDUCTION CAN BE ACHIEVED THROUGH EMPATHY, NOT SHAME

These flexible and pragmatic perspectives are an example of the *harm reduction model*, which has been successfully applied in various fields, especially public health, where it has been shown to improve social outcomes (Hawk et al., 2017). The harm reduction model recognizes that certain behaviors may be deeply rooted parts of people's lives and, instead of judging or punishing them, this approach aims to reduce the risks and consequences associated with those behaviors. Harm reduction works from a foundation of empathy, recognizing that people often struggle with behaviors that are not easy to change, or they may not have access to the resources and options that would support change. This model emphasizes avoiding shame or stigma toward individuals or groups while actively encouraging and celebrating incremental change.

Blame and shame rarely lead to lasting change because they can trigger defensive responses instead of fostering understanding or collaboration. When people feel blamed or shamed, they might dig in their heels, resist taking responsibility, and even withdraw completely. These emotional reactions can shut down productive conversations and hinder problem-solving, making it harder to tackle the root issues. The same goes for how we perceive ourselves. People who are committed to creating a better world find it easy to slip into negative self-talk and think "I'm not doing enough" or "I'm not making progress fast enough."

> *Blame and shame rarely lead to lasting change because they can trigger defensive responses instead of fostering understanding or collaboration.*

Let's take a closer look at whether harsh self-criticism is undermining your efforts. We need more than a sustainable world; we need a sustainable sense of self as well.

We might worry that by setting boundaries on our efforts we could be letting ourselves off the hook too easily, almost like a form of personal *greenwashing*: wanting to feel okay without making the required effort to be truly effective. But pushing ourselves too much can be problematic. This is true for anyone, and we can take a word of warning from the burnout commonly seen among people who work in environmental fields. Research conducted by the Oxford Brookes Business School and the Climate Change Coaches (2024) found that 62% of sustainability practitioners experienced burnout over the past year; in addition, 69% reported difficulty staying motivated, and only 53% felt that they received adequate support for their well-being.

We all need to learn ways to pace and replenish ourselves in order to remain present, healthy, and engaged for the long run. Let's explore some common ways that negative self-talk manifests as the voice of the inner critic that drains our energy and undermines our efforts and how we can address it.

- **Perfectionism.** You have already seen how perfectionism, one aspect of black-and-white thinking, can get in the way of worthwhile climate efforts: "That wasn't good enough," or "I'm not good enough." Remind yourself that mistakes are part of learning.
- **Imposter syndrome.** This can manifest as the belief that we need specialized training in order to really make an impact:

"I'm not qualified for this, and people will find out!" Remember that we're all on a learning curve, and you can grow your skills as you go.

- **Comparison with others.** Similar to imposter syndrome, this habit of mind automatically measures us against others. It may include a fear of being rejected or criticized: "Look how impressive they are. I'll never measure up." Return your focus to your unique journey and your own talents, path, and timing.

The key to overcoming negative self-talk, self-blame, and shame is patient practice. Over time, you can create a more compassionate, empowering, and realistic inner dialogue. It is important to be gentle with yourself and be aware that changing our patterns of thought is a gradual process. Try the following exercise to cultivate more empathy for yourself. It works best when practiced regularly.

Practice: Taming Your Inner Critic

Choose an experience related to your climate efforts during which you have noticed self-doubt or negative self-talk. Write down one of your critical beliefs. It may have many layers, but simplify it in a single sentence, for example, "I'm not good enough" or "That success doesn't count; it was just luck."

Now answer these questions about your judgmental statement:

1. Is this thought based on facts or on assumptions?

____

____

____

____

____

2. What evidence do I have that contradicts this thought?

____

____

____

____

____

3. How do I feel when I'm having that thought?

____

____

____

____

____

4. Are there certain behaviors—like binge-watching TV or compulsive shopping—that tend to take over when your critical thoughts get out of control?

____

____

____

____

____

Now reframe your thought from a kinder and more realistic perspective. For example, "I'm not doing enough" can be replaced with "I'm doing the best I can with the time and energy I have right now."

____

____

____

____

____

We are often kinder to others than we are to ourselves. We can use compassionate self-talk to empower ourselves to make gradual, realistic changes that make a difference and lead to additional positive changes. In the following exercise, you can practice compassionate self-talk by writing a letter to yourself.

Practice: Self-Compassion Letter

Expand the new and more caring view of yourself that you expressed in Practice: Taming Your Inner Critic by writing a self-compassion letter. Offer yourself words of encouragement, understanding, and empathy. Acknowledge the challenges, but remind yourself of your strengths and resilience. Some of us can have difficulty directing this kind of empathy to ourselves. If you need some help, begin the letter as though you are writing to a close friend who is struggling with self-blame or shame. You can also imagine that a friend is writing the letter to you.

____

____

____

____

____

Practice: Self-Compassion Letter

These practice exercises for transforming your inner critic can be applied to each example of negative self-talk that you become aware of. And it can't be said enough: They work best when practiced consistently.

TAKE HEART, AND REMEMBER THE BIG PICTURE

You have been looking closely at your own thoughts, feelings, and actions, and now it is just as important to zoom out and take in the wider landscape of climate work. Whether you are already deeply engaged or just beginning to explore your place in these efforts, remember that no one carries this alone. Change grows through shared effort, shared momentum, and the power of many people doing what they can over time. Let's now take a look at a few highlights from the past.

Early U.S. conservation efforts began with milestones like Yellowstone National Park in 1872, and were later advanced by figures like John Muir and Theodore Roosevelt, who helped establish the foundation for protecting natural landscapes (Runte, 2010). The 1960s and 1970s marked a major shift as Rachel Carson's (1962) book *Silent Spring* exposed the dangers of pesticides and energized the modern environmental movement (Lear, 1997). At the same time, the environmental justice movement revealed how marginalized communities, especially communities of color, were disproportionately harmed by pollution and industrial development (Bullard, 2000; Taylor, 2014). The first Earth Day, April 22, 1970, and key pieces of legislation, such as the Clean Air Act of 1963, the Federal Water Pollution Control Act of 1972 (known as the Clean Water Act), and the Endangered Species Act of 1973, reflected a rising national commitment to environmental protection (Human Rights Research Center, 2025).

The North Carolina PCB Protest of 1982 further highlighted environmental racism (Bullard, 1993), and global action expanded through agreements like the Montreal Protocol (in 1987) and the Rio Earth Summit (1992) led to the UN Framework Convention on Climate Change, the first international treaty committing nations to address climate change (United Nations, 1992). In 1991, the First National People of Color Environmental Leadership Summit produced the *Principles of Environmental Justice* (Climate Justice Alliance, n.d.), which affirm everyone's right to a healthy environment (Bullard, 2005).

By the 2000s and 2010s, climate change had become the central focus of environmental work, with global efforts strengthening through the Kyoto Protocol and the Paris Agreement (Bodansky, 2001; United Nations, 1997, 2015). Conservation approaches expanded to include sustainable development, renewable energy, and ecosystem restoration (Secretariat of the Convention on Biological Diversity, 2020). Today, rapid progress in solar and wind power (International Energy Agency, 2023b), energy storage technology (International Energy Agency, 2022), regenerative agriculture (Intergovernmental Panel on Climate Change, 2022), and the exploration of carbon-removal strategies (National Academies of Sciences, Engineering, and Medicine, 2019) is accelerating the transition to a low-carbon future and reshaping modern conservation efforts.

Although these efforts have not yet brought about the complete changes necessary for protecting the Earth, this long journey underscores the dedication of countless individuals who have worked to protect the environment. They have planted the deep roots required to ensure that constructive changes are anchored and sustainable. Just as a flower bulb quietly develops beneath the surface before breaking ground, social change often begins with unseen efforts that slowly build momentum before they become visible to the larger world. Once the flower bursts above ground, it now depends on sunlight for growth—much like social change relies on widespread public recognition and participation. This stage represents the visible success of all the invisible work that came before it.

At the same time, it is important to recognize that progress does not follow a steady or linear path. There are periods when the political climate supports environmentally beneficial changes and others when it stalls or even reverses that progress. Consider your relationship with the climate as a long-term commitment to environmental well-being that evolves through different seasons. Some will be full and demanding; others will call for rest, reflection, and recalibration. There will be times of exciting achievements and celebrations and times of disheartening setbacks. Sometimes, progress will be visible; other times it will slowly advance underground, quietly gathering strength until it is ready to emerge again. Passion is not about constant urgency or output; it is about sustainability. It respects boundaries. It makes space for rest—and, as you have been learning, it thrives in community.

Research shows that public support for climate action is at an all-time high (Leiserowitz et al., 2023), with a significant majority of people around the world pushing for stronger government initiatives to address climate change. The Peoples' Climate Vote 2024, headed by the UN Development Programme, surveyed more than 73,000 people across 77 countries and found that 80% of the respondents said they want their government to take more robust action on climate change (UN Development Programme & University of Oxford, 2024).

Behavioral researchers at several European universities also found that a large majority of the global population supports climate action and is willing to bear personal costs in order to fight climate change (Boneva et al., 2024). In fact, 86% of a group they surveyed reported that they follow pro-climate social norms, and 89% advocated for stronger political action.

This is how social change reaches a tipping point and becomes unstoppable, encouraging new laws to be passed, societal norms to shift, and policies to be enacted. Your role in shaping this future is a vital part of the story. Pause now and sense the momentum that has built from past efforts, from the community around you—and feel the resiliency within you as you look ahead and step onto new ground.

Forces of change move through communities and through each of us, inviting clarity, strength, and resilience, yet even as we step toward new possibilities we often carry grief for what is shifting, delayed, or disappearing. In Chapter 4, we explore one of the most profound emotional responses to living in a rapidly changing world: eco-grief—what it is, why it matters, and how to make room for it.

CHAPTER 4

PROCESSING ECO-GRIEF

Eco-grief is the deep ache that arises when the living world on which we depend is being harmed: when species disappear, landscapes transform, climate systems destabilize, and the cultural and ancestral places that help ground us are disrupted. Researchers have described this grief as a natural, highly meaningful response to ecological loss and a sign of emotional attunement rather than pathology (Cunsolo & Ellis, 2018). Eco-grief emerges because we are woven into the living world, not separate from it. When we feel eco-grief, it is often because something in us recognizes the depth of that relationship and the reality of its disruption.

Rather than bracing against these feelings—distracting ourselves, dulling our emotional responses, or pushing the pain aside—or trying to silence them altogether, in this chapter we will learn the benefits of skillfully opening ourselves to them. Eco-grief can sit beside our love, fear, longing, and hope, reminding us of all we value. Naming what we are feeling interrupts the avoidance and numbing that only deepen our distress, and it keeps our eyes and heart open to who and what matter most.

Eco-grief is profoundly relational. Carried alone, it can feel heavy, overwhelming, and isolating. But when it is shared in community, grief becomes something else: a thread that binds us, a communal recognition of what we are losing—and what we refuse to give up on. Research shows that unprocessed eco-grief can contribute to burnout and hopelessness, whereas acknowledging and making sense of difficult emotions supports psychological flexibility, social connection, and sustained engagement (Hickman et al., 2021; Pihkala, 2022). When shared, eco-grief can transform into solidarity, creativity, and a renewed sense of agency.

In this sense, eco-grief is both a painful rupture and a form of guidance. It exposes our vulnerability, but it also kindles our commitment to protect what we love. Approached with openness, eco-grief becomes an invitation to slow down, reach toward one another, and imagine new ways of living on a planet in flux. Rather than blocking action, it often becomes the quiet catalyst for meaningful and enduring change.

Before we dive into exploring our relationship to eco-grief, let's practice a resiliency exercise that will help you stay grounded in the face of the painful realities you will be reading about. As we learned in Chapter 2, building emotional resilience for addressing climate change is not about bypassing feelings or avoiding hardship but instead developing the ability to face difficult truths without shutting down or reacting impulsively. This helps us keep our hearts and minds open as we become more capable of responding effectively and responsibly.

Practice: Box Breathing

The box breathing technique, also known as *square breathing*, is a simple and effective practice for regulating the nervous system. To begin, sit comfortably upright, or lie down if you prefer. You can adjust the pace of the breath if four counts feel too short or long: The key is consistency. If you find it difficult to hold your breath at any point, ease into it gradually and do not force it. As you practice regularly, your lung capacity and comfort with the practice will grow.

Imagine tracing four sides of a square, with each side corresponding to a different step:

1. **Inhale:**—Breathe in through your nose for a count of four.
2. **Hold:**—Hold your breath for a count of four.
3. **Exhale:**—Exhale slowly through your mouth for a count of four.
4. **Hold:**—Hold your breath for a count of four.

Repeat this cycle several times, focusing on the breath, tracing the square, and keeping the counts consistent. After five or six breath cycles, release your focus on the practice and allow your breath to return to a natural rhythm.

How do you feel after doing box breathing? Take a moment to tune into your body, mind, and emotions, and write about what you notice.

____

____

____

____

____

Practice: Box Breathing can be done any time you need take a brief break and steady yourself while sitting with the painful and challenging realities of climate change.

FORMS OF ECO-GRIEF

As we explore experiences of eco-grief, we must remember that these are not just statistics; they represent the lives of children, parents and grandparents, animals, homes, cultures, communities, and livelihoods. Eco-grief is the flip side of our care and love: It reflects our emotional and empathic bond with life. It is what helps us connect and feel a sense of belonging.

> *Eco-grief is the flip side of our care and love: It reflects our emotional and empathic bond with life. It is what helps us connect and feel a sense of belonging.*

Eco-grief is a deep, complex, and prolonged emotional process that follows significant, sometimes layered, loss and can include a variety of feelings. We often link grief with sadness, but it is not a one-size-fits-all emotional experience: It is a spectrum of feelings that can ebb and flow, often unpredictably. Experiencing a mix of climate emotions at different times, or even all at once, is normal.

Climate change elicits many forms of grief, often experienced together, each tied to different facets of the polycrisis (Pihkala, 2024; see Chapter 1, this volume). Eco-grief is a response to changes in both the physical world as well as our emotional, cultural, and social well-being. Working through different types of eco-grief helps us process our experiences. This is a crucial step in healing that enables us to find meaningful ways to respond to the crisis through collective action.

As you read through the descriptions of different forms of eco-grief, pay attention to the ones that bring up strong emotions in you.

Disenfranchised Grief

Eco-grief is often not openly acknowledged, socially validated, or supported by others; instead, it sometimes is viewed as insignificant or an overreaction. This can make the person experiencing it feel isolated and misunderstood, as though their grief is not allowed to be fully experienced or expressed (Cunsolo, 2017). This type of grief can be harder to process because the person grieving may feel like they do not have permission to grieve or that they may be judged for their emotional experience.

Collective Grief

This type of grief is shared by communities or entire groups who are experiencing the direct impacts of climate change, for example, those displaced by rising sea levels, wildfires, or extreme weather events (Pihkala, 2022). It reflects the emotional toll of seeing one's community torn apart and culture disrupted.

Anticipatory Grief

Anticipatory grief occurs when people mourn the expected loss of something in the future, such as the worsening of climate change impacts or the knowledge that future generations will inherit a damaged planet (Randall, 2009; Ray, 2020). It is a sense of mourning for what has yet to come and is often accompanied by feelings of helplessness.

Historical Grief

Indigenous communities in particular may experience this form of grief, which has been created by a long history of losing ancestral lands and ways of life—first to colonialism and land theft, then to industrial degradation, and now to climate change, with each new generation carrying the accumulating layers of cultural trauma (Brave Heart, 1998; Lebel et al., 2022).

Transgenerational Grief

Transgenerational grief is passed down through generations and is often linked to the legacy of environmental harm caused by previous generations (Pihkala, 2022). Younger generations may grieve the world they inherited, and older generations mourn their complicity in the damage done during their lifetime (Swim et al., 2022). This can create tension and conflict within families (Ayalon & Roy, 2022).

Existential Grief

This is a deeper, philosophical grief that is tied to the broader existential questions about the very nature of humankind. As the polycrisis brings humanity to a crossroad, we are gripped by the question of whether we can muster the collective awareness to ensure the health of the Earth and our own survival (Van Susteren, 2018).

Personal Grief

For some individuals, climate change delivers a personal experience of multiple losses through the destruction of their homes, displacement, or the death of loved ones in climate-related disasters (Pihkala, 2020a). Although these are deeply personal experiences, they are also part of the larger context of a changing world.

Compassionate Grief

Compassionate grief occurs when people feel sorrow and loss for those affected by climate change, even if they are not personally experiencing it (Ray, 2020). Activists, environmentalists, climate scientists, journalists, and a broad range of frontline workers empathize with and grieve for those suffering because of climate disasters and environmental harm (Kelly et al., 2024; Pihkala, 2019).

Grief From Moral Injury

In the context of the climate crisis, *moral injury* refers to the psychological and emotional distress individuals experience when they feel they have violated their own ethical beliefs or values or they have been complicit in actions that contribute to harm being done to other humans, animals, and the environment, even if indirectly (Davenport, 2017). This can happen when people are faced with the overwhelming knowledge of the damage being done to the environment yet feel powerless to stop it—or when they realize that their actions (or inaction) are contributing to the ongoing crisis. Moral injury is a complex emotional challenge that can deeply harm our sense of identity and purpose. It highlights how the climate crisis is not just an environmental issue but a deeply ethical one as well.

Grief From Loss of Future Security

As climate change worsens, many people, espccially younger generations, grapple with a profound sense of loss regarding their future (Hickman et al., 2021). They often feel as though their potential futures have been stolen from them, leaving them to mourn the uncertainty and instability that seem to loom ahead. It is a very real loss, even if it has not yet materialized in a concrete way.

Ambiguous Grief

Ambiguous loss complicates the grieving process because it does not fit neatly into societal expectations of what loss looks like, and it leaves people with unresolved emotions and a sense of disorientation. This form of eco-grief is often experienced by individuals or communities who are living through gradual, ongoing environmental changes that make it impossible to pinpoint a precise moment of loss, even though they feel its emotional weight every day (Cunsolo & Landman, 2017).

PROCESSING ECO-GRIEF

Let's take some time to apply what we are learning to your own eco-grief. Eco-grief opens us to deeper parts of ourselves, reconnecting us with our hearts and our place in the web of life. It can become a powerful catalyst for transformation. A good first step in working with eco-grief is to acknowledge the unprecedented and frightening scope of climate change while understanding that acceptance does not mean resignation. Acceptance helps us convert any energy that goes toward psychological denial or distraction to be released and redirected to build a more grounded, peaceful, and proactive mindset.

> *Eco-grief opens us to deeper parts of ourselves, reconnecting us with our hearts and our place in the web of life.*

There is no right or wrong way to grieve, and every person's journey is unique. As you proceed through the next three practice exercises for processing eco-grief, remembering that grief is multidimensional can help normalize the complexity of these feelings and create more space for healing.

Practice: Identifying the Nature of Your Grief

Reflect on the types of eco-grief described in this chapter, and identify two or three that resonate most strongly with you. Write about them—specifically, about the experiences or information that trigger your feelings. What are the strongest emotions? Where do you feel them in your body? What thoughts arise alongside these feelings? Resist the urge to jump into problem-solving or action just yet. Allow yourself the space to simply be with the feelings. Remind yourself that it's okay to feel whatever comes up, without judging your emotions or the intensity of your grief.

____

____

____

____

____

____

____

Practice: Identifying the Nature of Your Grief

Now let's try some somatic grief practice. *Somatic practice* is any activity that helps you connect with and become more aware of your body and its sensations. It focuses on how your body feels, rather than just your thoughts. Grief is often experienced as a deeply physical sensation because the body and nervous system hold memories of the pain associated with loss. Grief can show up as tension, discomfort, or other physical sensations, including a sense of tightness in the chest, a lump in the throat, agitation, or fatigue. Somatic practice will help you become more aware of these physical responses and use them to work with your eco-grief more deeply while staying grounded in the present moment.

> *Grief is often experienced as a deeply physical sensation because the body and nervous system hold memories of the pain associated with loss.*

If this approach is new to you, it might feel uncomfortable or awkward at first. Consider this an opportunity to step into your stretch zone and become more at ease with discomfort, finding reassurance that these practices are proven therapeutic tools for processing grief (the stretch zone is discussed in more detail in Chapter 5). At the same time, if you feel the need to modify the practice on the basis of your own sense of self-care, feel free to make adjustments, or stop the practice and transition to journaling, a grounding exercise, or another self-care activity that feels supportive.

Practice: Processing Your Somatic Grief

This practice exercise is helpful for many, but persons with unresolved trauma may notice periods of emotional intensity or dysregulation. Although intense feelings are not a sign that anything is wrong, you may find it helpful to work with a therapist or counselor when exploring deeper emotional material. If possible, consider having someone you trust available for support before beginning. Resources for finding a climate-informed therapist can be found in the Appendix at the back of the book.

You can do this practice sitting up or lying down.

1. **Create a safe space.** Find a quiet, comfortable place where you will feel safe and undisturbed for at least half an hour. You may want to light a candle or have a soft blanket nearby: Bring whatever helps you feel comfortable and grounded into your practice space.
2. **Set an intention.** Choose one of the grief experiences you wrote about in Practice: Identifying the Nature of Your Grief, and focus on those feelings, building an emotional connection to what you are grieving. You may want to close your eyes as you call to mind the source of your grief. Give yourself permission to simply notice what you are experiencing with kind curiosity. You might wish to set an intention, such as "I'm honoring my grief and letting it move through me."
3. **Breathe and tune into your heart.** Take three full, intentional breaths, feeling how the breath moves in your body and noticing the letting-go quality of the exhalation. Grief is often felt deeply in the heart's center. Bring your awareness inside to the center of your chest. You may want to try resting your hand gently over your heart to help connect to that area. Breathe into your hand; breathe into your heart for several cycles. Notice anything you feel.
4. **Tune in to your body and breathe into the grief.** Let your awareness gently move through the rest of your body, noticing where grief shows up physically. You might feel it as a tightness in your chest, a heaviness in your shoulders, tension in your stomach, or other sensations in different parts of your body. Choose the area where the sensation feels strongest today, trusting that you can return to explore other areas another time.
   - At first, do not try to change anything; simply observe how it feels. Stay with this feeling, giving it space to reveal itself, without rushing to analyze or fix it.

- Now allow your breath to flow into those areas of tension. When you inhale, imagine that you are breathing into the areas where you feel weight, numbness, or tension, honoring the presence of grief. On the exhale, help those areas soften and receive the gentle vitality of your breath.

5. **Respond with your body.** At times, grief needs to be expressed through movement or other sensory experiences, such as a gesture, a sound, or an image. You might feel an urge to rock, shake, stretch, hum, sigh, or just be still. For example, if you notice tension in your shoulders, try gently rolling them or stretching softly. You may also feel drawn to placing your hand over areas of your body where the sensations are strongest, offering yourself a gentle, compassionate touch. Trust your body's impulses without overthinking or trying to control them: There are no wrong actions, just your body expressing what it needs to release.
6. **Respond using imagery.** Sometimes, grief may feel stuck and show up as a distinct physical impression or image, for example, a heavy rock or a tight metal band. If this happens, approach it with tenderness and curiosity. Explore its size, shape, texture, and color, seeing it as a messenger of grief rather than a problem. Notice what this image might need—perhaps light, touch, or breath—and imagine bringing that remedy to it. With your gentle attention, it may begin to release on its own.
7. **Close the practice.** Once you have allowed yourself to fully feel your grief, take a moment to thank yourself for letting it move through you. You can also do something that offers you comfort, such as placing a weighted blanket or cushion on your body. Acknowledge that you are bringing this practice to a close for now. To transition, spend a few moments grounding yourself by placing your hands on the Earth or sitting or standing with your feet firmly on the ground. If you are outside, that's great! Even indoors, you can still feel or sense the grounding energy of the Earth beneath you.

Practices like this can bring up unexpected emotions or sensations. Whatever you are experiencing is all part of the journey. As you have seen, writing can be a powerful way to integrate what you have experienced. Write about the sensations you felt in your body, the emotions that surfaced, and any messages or insights that arose.

____

____

____

____

____

____

____

____

Personalizing your somatic grief practice can help you connect with your grief in ways that feel most authentic to you, and the format you have tried today can be shaped with additional elements. For example, if you find box breathing (Practice: Box Breathing) or balanced breathing (Practice: Balanced Breathing) particularly helpful, you can add them at the beginning or end of your somatic grief practice. If you enjoy music, drawing, or collage, those are also great way to integrate your experiences in addition to writing.

Grief has its own timeline, and honoring it as it arises is key to processing it. Recognize your climate grief without trying to force it to go away or pretending it does not exist. Continue to create a time and place where that grief has a place to express itself physically and emotionally. This is an ongoing process, and returning to it again and again, with patience and self-compassion, is how you grow more skilled at living with what is challenging.

Practice: Creating a Remembrance Space

A remembrance space offers a tangible way to stay connected to both grief and inspiration. It serves as a physical tribute and emotional link to people, places, and things you have lost while also honoring your deepest eco-conscious values and the beauty of the Earth that deserves protection.

Find a space on a mantel, small table, corner of a desk, or any place where you can assemble a few meaningful objects. Choose a location that invites reflection and contemplation. The selection of items you place there is based entirely on what you find personally significant. The act of creating and tending to the remembrance space can become a mindfulness practice that helps you stay connected to your grief and your inspiration for climate action while also gradually supporting your healing.

You could consider including items from these categories:

- The elements: a candle; small bowl of water; a plant; or a natural object, like a shell or leaf
- Personal possessions: photographs or drawings of a person or place, perhaps a beloved object, like an old toy or a piece of jewelry
- Images or symbols: crystals, icons, seeds, or figurines that hold meaning for you
- Notes or quotes: messages or phrases that express feelings or even unfinished conversations

Given the ever-shifting nature of climate change and your personal healing journey, what you want reflected in the remembrance space will likely evolve over time as you add and remove items. Embrace this as a living process, a meaningful expression that stays aligned with your ongoing growth. With climate impacts currently on an escalating trajectory, your skillfulness at attending to your grief will be a strength and an essential element for navigating the future with resilience and purpose.

THE COLLECTIVE EXPRESSION OF GRIEF

Although this book is designed to help individuals navigate their climate-related emotions, the practice exercises can also be shared as powerful tools for collective expression. In many traditional cultures grief is seen as something meant to be experienced with others, not in isolation. Shared mourning fosters connection, understanding, and mutual support during times of deep sorrow. It supports communities to honor what was lost, build cohesiveness, and move forward together.

Creating a community remembrance space is a powerful and evocative group practice in which each person describes the significance of the item they add to the group arrangement. The practice of breathing into grief can be done in a group and followed by the sharing of experiences. Imagine bringing these practices into various gatherings: a work-related group, a classroom, a meeting with close friends, and more.

Up to this point, we have been exploring the emotions we can feel directly, the ones that show up as tension, fear, urgency, grief, hope, or motivation. But our emotional lives are shaped by more than what rises to our conscious awareness. Beneath the surface lie patterns of thought, interpretation, and reactions that quietly fuel our feelings, our actions (and inaction), and the beliefs we hold about what is possible.

In Chapter 5, we turn our attention to these deeper layers—to how some eco-emotions may originate from processes that are rooted in the architecture of our evolutionary brain and operate beneath our conscious awareness. Exploring this inner terrain can reveal surprisingly fertile ground for change and help lay the foundation for creating a more climate-safe future.

To begin, we will look at why evolution equipped us with certain error-prone mechanisms for processing information that helped our ancestors survive but can work against us in today's rapidly changing modern world. Is there any adaptive value to having mental shortcuts that sometimes distort reality, especially when those distortions can lead to mistaken perceptions or even large-scale environmental harm? In Chapter 5, we explore *cognitive biases*: patterns of thinking that often help us get through daily life but can become obstacles when we are trying to understand and respond to the vast, complex challenges of climate change.

CHAPTER 5

UNDERSTANDING COGNITIVE BIASES THAT CONTRIBUTE TO CLIMATE CHANGE

Consider that many scientific facts about climate change are now widely known and no longer limited to journals and conferences yet, despite the decades that have been spent building scientific clarity, the facts alone have not spurred the sweeping changes required to slow or reverse global warming. Why? One of the most fundamental explanations comes not from the science but from how our minds react to it. As we absorb new information, it is filtered through long-standing mental habits and cultural storylines. These patterns are rooted in our evolutionary wiring and the social worlds in which we live, and they can make a slow-moving global crisis feel strangely abstract, even when the evidence is right in front of us (Begum et al., 2025; Stoknes, 2015). In this sense, we resemble the proverbial fish with no concept of water: We move through our assumptions and default responses so automatically that we barely notice the currents steering us. If you have ever caught yourself drifting through a page of alarming headlines in a kind of emotional trance—taking in the words but not actually feeling them—you have experienced the bias described as *psychic numbing* (Slovic, 2007).

Alongside our emotional responses, beneath our conscious awareness, our minds are continuously sifting through an enormous mass of incoming information, making rapid judgments about that information, and automatically filling in gaps. The result is a vast storehouse of mental shortcuts known as *cognitive biases* that shape how we interpret just about everything, including climate change and how we assess its risks and organize our approach to it. Even if you are already fully awake to the realities of a warming world, these biases still operate beneath the surface until we build the awareness to notice and work with them. In this chapter, you will learn how to recognize several common biases that can undermine your efforts to confront climate change wisely, along with practical ways to stay alert to other biases as they arise and bring them into conscious awareness so you can reflect on them rather than be steered by them.

WHY COGNITIVE BIASES MATTER IN FACING CLIMATE CHANGE

As you have discovered, our climate emotions are genuine signals of our deep connection to the living world. However, if we explore only the emotional landscape, without understanding the underlying thought patterns that accompany it, we will miss half the picture. We often imagine our brains working like powerful computers, taking in facts and automatically generating accurate, rational conclusions, but below the surface far more is happening. For example, with the *present bias*, we might put off preparing for extreme weather because today feels calm and the future feels far away, even though we know stronger storms are coming (O'Donoghue & Rabin, 1999).

Cognitive biases help us interpret information and make decisions quickly, but they are likely to distort our perception of existential realities, such as climate change and social justice issues, without us realizing it (Tversky & Kahneman, 1974). When a cognitive bias is triggered, we may fall back on interpretations that feel familiar rather than tracking down more information to get a complete picture. For example, we hear about a climate disaster on the other side of the country, and our *negativity bias* might make us see that disaster as proof that nothing we have done matters, leaving us feeling discouraged and defeated (Baumeister et al., 2001).

The good news is that once we learn to recognize these tendencies they can be transformed from unconscious habits into conscious insights. First, it is crucial that we learn to move from automatic thinking toward awareness: When we detect a bias in our thoughts, we can pause, reflect, and choose a different response, rather than being swept along by default thinking. For example, if someone makes a point about climate that contradicts what you believe, your first instinct might be to dismiss it. That is a classic moment of *confirmation bias*: the tendency to favor information that supports what one already believes while discounting or ignoring anything that challenges it (Nickerson, 1998). But if you pause, notice the reaction, and get curious instead—"Tell me more about how you see it"—you can engage in the kind of genuine dialogue you will practice in Chapter 6.

In the context of climate change, this awareness matters profoundly: If we notice that a bias is narrowing our sense of possibility, minimizing risks, or amplifying our fear, we can recalibrate our thinking in ways that support clearer judgment, emotional steadiness, and more effective action. By bringing both our emotional and cognitive lives into awareness, we will build a more integrated and grounded approach to climate engagement.

Researchers and cognitive scientists have catalogued well over 170 distinct cognitive biases (Benson, 2016; Caverni et al., 1990). That's why for us fish, understanding our mental and cultural water is so essential. When we learn to recognize the forces that are unconsciously manipulating our attention, our understanding of a situation, and our sense of urgency, we will gain access to new choices.

SHORTCUTS AND BLIND SPOTS

Here is a story that is frequently used to illustrate how our biases operate: A father and son are in a car accident, and the father is killed instantly. The boy is knocked unconscious, but he survives and is rushed to the hospital for immediate surgery. The doctor enters the emergency room, sees the boy, and exclaims, "I can't operate on this boy—he's my son!"

You may have figured out that the doctor in the story was the boy's mother. Most men and women automatically assume that the emergency room doctor is male, and that is a classic example of *gender bias* (Moss-Racusin et al., 2012). It might have only taken you a few moments to think through this problem, but in everyday life we may never discover our biases until an enigma like this makes us pause and consider.

Not all cognitive biases are detrimental; in fact, they can be very effective in helping us react fast to immediate needs and threats. They quickly filter incoming information and direct our attention to what is most pressing in the moment. Throughout much of our evolution, it was crucial to focus on urgent threats: what could harm or kill us right now rather than later. Faced with frequent danger in the form of animals and marauding tribes, our ancestors developed an orientation that emphasized paying attention to immediate risks and opportunities and overlooking more distant, complex, and abstract vulnerabilities. This adaptive tendency, known as *hyperbolic discounting* (Ainslie, 2001), has roots that extend deep into human evolution and has been shaped across hundreds of thousands to millions of years. It helps explain why we struggle to grasp the significance of climate change when it presents as statistical trends and long-term gradual shifts.

Few people on this planet want to harm the Earth: The challenge is that our immediate priorities often overshadow crises with longer time horizons. In this chapter, we explore other biases that hinder our progress in tackling climate change and discover how we can navigate them to promote a more eco-conscious perspective with a broader and more accurate understanding of our world. As you learn about biases, I encourage you to pause and look deeply within, using the guided discernment practices detailed in the practice exercises to reflect on how you might recognize the bias manifesting in yourself, at work, and in your relationships with friends and family.

Kristin Pauker, a psychology researcher at the University of Hawaii, explains that biases are "sticky" (Madhusoodanan, 2023, para. 3)—meaning they can be surprisingly hard to change—because they are deeply rooted in our evolutionary history and early childhood experiences. Although some biases may appear reasonable and have a kernel of truth, the aim of discernment is to carefully analyze these patterns, preserving the truth while correcting the distortions. In doing so, we gain a clearer understanding of how biases hinder our engagement with climate issues, which empowers us to make the adjustments needed to better align our actions with our values.

THE BYSTANDER BIAS

The term *bystander bias* describes our tendency to believe that someone else will take care of a crisis, and this instinct has served a practical purpose in our past (Myers & Twenge, 2019). In our hunter–gatherer days, if a dangerous animal imperiled the group, it wasn't efficient for everyone to respond: That would just put more people at risk. Roles were clearly defined in smaller human groups, and the appropriate person would automatically address their designated task or threat. However, this mindset has led us to believe that our leaders will take adequate action to address climate change—even though many of us are now aware that government initiatives have fallen tragically short in terms of the scale and urgency required to heal this escalating crisis. And as social groups grow larger, the bystander bias becomes even more pronounced.

I've heard many people say things like, "I'm not a climate scientist, engineer, grassroots leader, billionaire, or politician . . ." or whatever roles they believe are responsible for managing the solution. This mindset assumes that someone in one of those roles will fix the problem and thus allows us to avoid responsibility for engaging in climate action. It is an internal shrug paired with the climate emotion of helplessness: "There's nothing I can do about it anyway." However, when we apply discernment to the problem we see that given the scope and complexity of climate change, everyone has a vital role to play. We need people from all walks of life—education, health care, parenting, farming, Indigenous leadership, civic positions, and beyond—to take responsibility for their own inner healing and then contribute within their areas of influence. Each person can find a meaningful way to engage using their unique talents and passions to help drive systemic change.

Practice: Transforming Your Bystander Bias

Take a moment to consider the following, and write down your thoughts and feelings as you explore these questions:

1. Where do you see the bystander bias in action in yourself or others with respect to climate change?

 ____

 ____

 ____

 ____

 ____

2. Do you see any benefits to standing aside? Why might you choose to avoid getting more involved?

____

____

____

____

____

3. What emotions come up as you reflect on this?

____

____

____

____

____

Now let's explore ways to deepen and increase your involvement in creating meaningful change.

1. List three actions, behaviors, or perspectives you are not currently practicing that could make a positive impact on the environment. Or, if you want to focus on how you are already involved, focus on this question: How can you expand or deepen those efforts? Examples include joining a community action nonprofit, choosing to increase the number of conversations you have about climate change, finding a community for emotional support, and leading an initiative at school or work.

____

____

____

____

____

2. Out of those three actions, behaviors, or perspectives, identify the one you believe is likely to have the most impact.

____

____

____

____

____

3. For the activity with the most impact, list three specific things you can do to launch the process or expand or deepen the impact of what you are already doing.

____

____

____

____

____

4. Who or what benefits from your action plan, and who or what is left out? Make any necessary adjustments as you consider this question.

____

____

____

____

____

5. From your three action items, identify the one that makes the most sense to start with (or the one that is the biggest priority). Are there any important reasons not to start with the item with

the highest priority? If so, choose an alternative that is more realistic to begin now and write about it.

____
____
____
____
____

6. What are some of the benefits if you follow through? Write about how you will feel if you take action.

____
____
____
____
____

THE SUNK-COST BIAS

Because of what is called the *sunk-cost bias*, we tend to persist in a certain course of action even when it leads to negative outcomes (Arkes & Blumer, 1985). That is because the more time, energy, or resources we have invested in a goal, the harder it becomes to abandon it, regardless of whether it is still the best choice. This bias also compels us to justify our actions because we believe that we shouldn't waste those efforts.

A classic example of a sunk-cost bias applied to sustainability investments is when a company continues to pour significant resources into a renewable energy project that is underperforming. This occurs simply because the company has already invested considerable time and money, even though more promising alternatives with better potential returns have become available.

It might seem logical to stick with our current path and hope that any problems will be fixed and the losses eventually recovered. However, this impulse often defies reasonable odds. It's like a gambler who has already lost everything but decides to place one more bet that they are convinced will magically win it all back.

Addressing the biases at play in your life requires a deep level of self-reflection. It often involves examining and shifting your priorities, patterns, and habits, which can be a challenging process. There can also be very real practical factors at play, such as the start-up costs involved in taking a new direction. These may include the time, expense, and energy required to learn new skills; the effort of renegotiating roles or routines; or the emotional cost of tolerating uncertainty and temporary discomfort while something new takes shape. When these initial costs feel high, it can reinforce the pull of staying with what is familiar, even when it is no longer serving us.

Practice: Transforming Your Sunk-Cost Bias

Look back at Practice: Transforming Your Bystander Bias. Let's take things a little further.

1. Think about yourself and your life situation and list three potential obstacles to moving forward with your new climate action. Are any of these obstacles created by the sunk-cost bias? In other words, are you so invested or preoccupied elsewhere that you don't have the time or energy to begin your new action?

   ____

   ____

   ____

   ____

   ____

2. What kind of support would help you find the time and resources you need to follow though and then stay on track? Be specific.

____
____
____
____
____
____
____

3. Where can you get that kind of support? Are you willing to develop more support now? If so, include a reasonable timeline for your next steps.

____
____
____
____
____
____
____

CONFIRMATION BIAS, LOSS-AVERSION BIAS, AND OPTIMISM BIAS

Biases rarely show up alone: They tend to interact and reinforce each other, creating patterns of thinking that can feel very automatic. For example, when we slip into the sunk-cost bias we are more likely to lean on *confirmation bias*, too, seeking out information that supports the choices we have already made and filtering out anything that might challenge them (Nickerson, 1998).

Loss-aversion bias can make this pattern even stronger. Because we are generally more sensitive to possible losses than potential gains, we often avoid changes that feel costly in the moment, even if they could benefit us later (Kahneman & Tversky, 1979). And *optimism bias* can give us an overly hopeful sense that things will work out fine, even when the evidence points in a different direction (Sharot, 2011). This shows up in our climate perceptions as well. For example, many people assume that their own community is less vulnerable to climate impacts than it actually is—even when local data suggest otherwise—because believing "It won't happen here" reassures us and calms our anxiety. These reactions are not about personal failings; they are reflections of how the human brain handles risk, uncertainty, and change.

What can help us move beyond these distortions? Kahneman (2011) wrote that our minds are most responsive to information that feels certain and immediate. When something is uncertain—for example, exactly how fast the climate will warm, or how severe future storms will be—it becomes harder to act on the facts we are certain about. You can make a difference, and so can policymakers and climate communicators, by helping make climate impacts feel more concrete and relevant: how exactly they are affecting our health, economy, neighborhoods, and the people and places we care about (Gifford, 2011). For example, if you live near the ocean, instead of talking about the rise in sea levels in general terms, you might point out how higher tides already flood a nearby road. Or with the increase in wildfires, you could share how smoke is affecting your child's school days. These everyday examples help translate abstract risks into real-world impacts people can relate to.

Research supports this. A study conducted by researchers at the University of Amsterdam found that we often underestimate the urgency of climate change or downplay scientific evidence because the issue feels distant and not directly connected to our daily life (Većkalov et al., 2024). When you can recognize how climate change touches your life—your routines, your community, your sources of meaning—and pair that awareness with actions that feel doable, it is much easier to stay engaged and sustain your efforts over time. For example, you might notice how repeated power outages during heat waves are disrupting daily routines, or how extreme heat leaves parks and playgrounds deserted by midafternoon. When those changes land as part of your lived experience, you can take a meaningful step in response, like helping organize neighborhood cooling stations, advocating for heat-safe schools and shaded public spaces, supporting farmers in adopting climate-resilient practices, or joining a local resilience group pushing for stronger community protections.

MAXIMIZE THE PROS, MINIMIZE THE CONS

Cognitive biases can pull us in both helpful and unhelpful directions and even offer advantages and drawbacks at the same time. Take confirmation bias, for example: It can be beneficial to have a supportive community where you can share ideas with like-minded people and deepen your understanding of topics you value. But this community can also lead to the formation of a social echo chamber that can intensify societal divisions and reduce your own critical thinking and questioning of misinformation (Nickerson, 1998). Once you recognize biases (e.g., loss aversion and optimism) in yourself, others, and society, you can start working with them instead of being unconsciously maneuvered by them.

Practice: Identifying the Pros and Cons of a Bias

Write about the pros and cons of the loss-aversion bias and the optimism bias, noting whether you recognize them operating in yourself, others, and society.

____

____

____

____

____

To maximize the pros and minimize the cons, you must slow down enough to notice when these biases are active and then balance them with a counter perspective. For example, if you feel pulled by loss aversion, ask yourself what you might gain by letting something go or trying something new; if the optimism bias is running the show, pair your hope with practical planning and what-if thinking. Add structure, perhaps in the form of checklists, small experiments, and accountability partners, to keep your optimism grounded and loss aversion flexible. And check your thinking through a third-person lens ("What would I tell a friend who is in this same situation?").

> *To maximize the pros and minimize the cons, you must slow down enough to notice when these biases are active and then balance them with a counter perspective.*

These simple practices help turn our biases into tools for making wiser, more balanced choices rather than obstacles that can get in our way. Understanding cognitive biases can also help us be more compassionate, because it reminds us that everyone's brains, ours included, take automatic shortcuts that can distort how we see the world. Instead of assuming that people are being intentionally difficult or irrational, we can recognize that they may be working with incomplete information, old patterns, or stress-induced thinking. When we see how easily we can misinterpret things it is much easier to meet others with patience and curiosity.

THE STRETCH ZONE

Working with our cognitive biases inevitably nudges us into the *stretch zone*: that place just beyond our familiar patterns where growth becomes possible. When we gently lean into that edge, we build psychological flexibility, expand our capacity to see more clearly, and strengthen our ability to respond rather than react. Questioning assumptions we have held for years is not always comfortable, but in the terrain of climate work our willingness to feel slightly off-balance is an unexpected ally: It strengthens our adaptability and helps us grow into new perspectives in unknown territory.

You have been learning how our unconscious biases interfere with effective climate action. Instead of trying to tackle each of the more than 100 identified cognitive biases one by one, you can practice stepping out of your comfort zone, just beyond your current abilities, and into your stretch zone where true learning happens. This approach helps bring your attention to patterns and encourages your growth, and it will likely touch many of the unconscious biases at the roots of your habitual thinking and behavior. We live in a world where many shy away from discomfort and awkwardness, but you will learn that it's okay to feel uncomfortable or uncertain about what you think, do, or say. Our capacity to stay present with discomfort is essential to both personal growth and collective liberation.

Entering the stretch zone requires a personalized approach. For example, someone who tends to dominate conversations in a group can stretch by pausing and allowing others to contribute before speaking. On the other hand, someone who is hesitant to speak up might challenge themselves to more actively share their perspective. Both actions create discomfort because they push against each person's natural tendencies, but both provide significant opportunities for growth. For the person who pauses, there is much to be gained from listening to others, and it fosters a more inclusive group dynamic. For the person who speaks up, it brings diverse viewpoints to the group, strengthens connections, and boosts the speaker's self-confidence.

Choose a new goal that challenges you to move outside your comfort zone in a way that supports both your climate efforts and personal growth. For example, if you are someone who avoids asking for help because you fear rejection or appearing weak, challenge yourself to ask for feedback or assistance, whether it is in a work context or your personal life. If you avoid group settings, try dropping into a community climate meeting. If you are someone who has trouble saying "no" because of a fear of letting others down, understand that often leads to overcommitment and eventual burnout. You could practice setting boundaries by saying "no" in small, low-pressure situations, such as declining an invitation or a request to perform a small task.

Practice: Identifying a Stretch Goal

Write about a new stretch goal. Be specific, and include when and where you can put this goal into practice in the near future.

____

____

____

____

____

TRANSFORM YOUR BIASES, LIVE WITH INTENTION, MAKE A DIFFERENCE!

As you have been learning, transforming your mind and emotions goes beyond simply coping with the eco-emotions triggered by the climate crisis. As illustrated in Figure 5.1, our eco-distress is, in fact,

FIGURE 5.1. Cognitive Biases in a Climate Feedback Loop

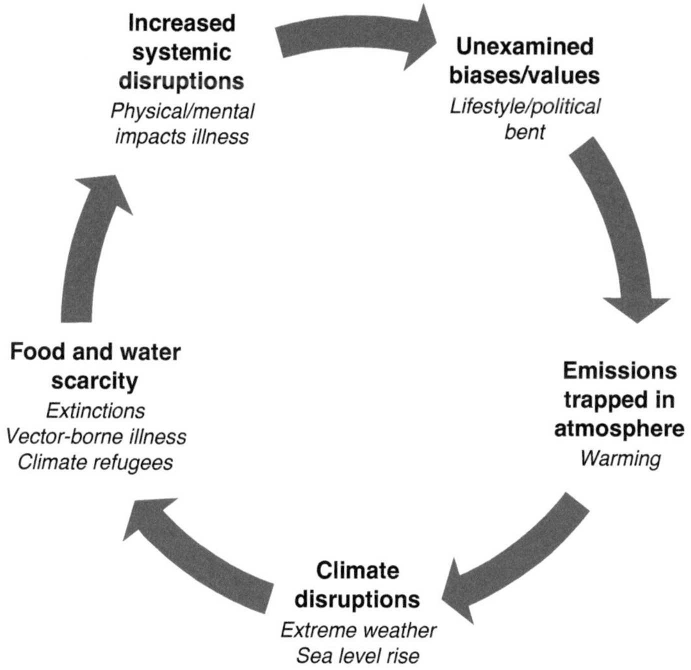

Note. Unexamined beliefs and emotional biases can create high-carbon behaviors, intensifying climate impacts and social instability, which in turn amplify systemic instability and psychological distress.

a reaction that is deeply entwined with the collective patterns and behaviors that drive us to unconsciously contribute to climate change and reduce our ability to engage effectively. Most explanations of climate change overlook the psychological causes and impacts and rarely consider the first element in this model, unexamined biases/values (at the 2:00 position in Figure 5.1), despite its significance.

The profound psychological effort you make to transform your biases will lead to living with more intention, meaning, and purpose, and it will enhance your ability to navigate the complexities that the changing climate brings. The opportunities to protect what we love on this beautiful Earth are innumerable, especially as we build our resilience and approach engagement with bold creativity.

> *The opportunities to protect what we love on this beautiful Earth are innumerable, especially as we build our resilience and approach engagement with bold creativity.*

Let's dive into an example of one young activist who discovered her passion and emotional strength to forge a path in defense of the precious beauty of her homeland—and has since inspired others worldwide.

Xiye Bastida, a young member of the Otomi-Toltec Indigenous community, grew up in the small town of San Pedro Tultepec, Mexico (One Earth, n.d.). Her connection to climate change is deeply personal: She saw first hand how her family and their land were devastated by environmental degradation. Illegal logging caused severe deforestation, which led to soil erosion and reduced the land's ability to retain water, ultimately decreasing agricultural productivity—patterns consistent with broader research on deforestation and climate-related ecosystem stress across Mexico (Intergovernmental Panel on Climate Change, 2022). Shifting rainfall patterns linked to climate change made the forest ecosystems more vulnerable to pests, resulting in widespread tree loss and disrupting local biodiversity and forest health (Intergovernmental Science–Policy Platform on Biodiversity and Ecosystem Services, 2019). In 2015, her parents secured jobs at the Center for Earth Ethics, and her family moved to New York City (Widdicombe, 2021).

Xiye felt overwhelmed by the enormity of the climate crisis (Zomorodi, 2020). As an Indigenous person, she has deeply held values that prioritize protecting the Earth, reflecting broader Indigenous worldviews centered on reciprocity, relationality, and responsibility toward land and future generations (Kimmerer, 2013). Xiye has spoken candidly about the weight of climate activism, describing a period when she was exhausting herself by running to meetings, sleeping 4 hours a night, getting on calls with Germany in the middle of the night, "because I thought that the world was on my shoulders" (Zomorodi, 2020, para. 14). But in New York, Xiye became involved with youth-led climate movements, joining groups like Fridays for Future, the global climate-strike initiative started by Greta Thunberg (https://fridaysforfuture.org/). Youth-driven climate mobilization has become a major force in global climate politics, demonstrating how young people exercise moral leadership and reframe climate change as an urgent justice issue (Fisher & Nasrin, 2021).

Xiye began speaking out at rallies, advocating for stronger climate policies and pushing for action on behalf of marginalized communities (One Earth, n.d.), which are disproportionately affected by climate change—a disparity widely documented in environmental justice research (Harlan et al., 2015). In February 2018, Xiye attended the 9th United Nations World Urban Forum in Malaysia, her first international conference where she spoke on a panel about Indigenous cosmology and its role in building resilient cities (Mejia, 2021). That same year, her involvement earned her the Spirit of the United Nations Award (One Earth, n.d.). In 2021, she was the only youth speaker invited to President Biden's Leaders' Summit on Climate, where she addressed 40 heads of state (Mejia, 2021). She has since been recognized for her work as the cofounder and executive director of the Re-Earth Initiative, an international youth-led organization that empowers front-line children and teens in 27 countries (https://reearthin.org/).

Although not all of us will find our work on an international stage, Xiye's journey is a powerful reminder that overcoming self-doubt and questioning our ability to make a difference often come from embracing one's true self, connecting with one's community, and turning personal experiences into a collective drive for change. Xiye's achievements are especially inspirational given that her family's experiences reflect how climate change affects people unequally because of reduced access to resources and opportunities—a pattern consistently identified in climate-justice scholarship (Thomas et al., 2019).

Experiment with taking risks, even when failure is a possibility—and don't forget to celebrate the small wins along the way. Embrace trial and error as an opportunity to learn from your experiences and build your social and emotional resilience. Be sure to breathe: Apply what you have been learning about staying within your window of tolerance as you explore ways to stretch your capacities.

How we approach social justice and transformative change—what qualities we bring to the table—plays a vital role in fostering true, transformative change. One key insight comes in the form of a paradox we explore in the next chapter: the value of slowing down in the face of climate urgency.

CHAPTER 6

SLOWING DOWN TO CLARIFY YOUR VISION

It is natural to think that, given the urgency of the climate crisis, the answer is simply to act faster, accomplish more, and overhaul our behavior immediately. Although momentous change is unquestionably required, the scale and complexity of environmental collapse have shown that it is not something we can fix through rapid-fire action alone. Hard scientific facts about fossil fuels have not been enough to move us, and cognitive biases are only part of the story. What is truly necessary is a deeper transformation, one that touches our sense of belonging, our imagination, and our perceptions, as well as how we embody our relationship with the living world. To begin that kind of shift, we need to learn how to focus our attention in deliberate, grounded ways that help us see more clearly and act wisely.

EXISTENTIAL INQUIRY

Let's begin with a reflective writing practice that invites you to explore and clarify your thinking about climate dilemmas on an existential level. This is an opportunity to step back from urgency and sit with the deeper questions: what it means to be human in a time of both unraveling and possibility, how we understand our place in the web of life, and what responsibilities and freedoms emerge from that understanding. By slowing down enough to ponder these layers, you will create room for insight, perspective, and a more grounded sense of direction.

There are no right or wrong answers in this first practice exercise; instead, the questions are designed to encourage a deeper reflection on your views of yourself and all of life in order to uncover hidden truths and lay the groundwork for insights. Take time with each question and then write your response, even (or especially) if you are unsure of your stance on the topic.

Practice: Deep Questions in a Changing World

1. **What does it mean to be human in a time of environmental collapse?** How do we reconcile our species' potential for innovation and progress with our role in causing the destruction of the planet's ecosystems?

   ___
   ___
   ___
   ___
   ___
   ___
   ___

2. **How do we define a meaningful life in the context of this global crisis?** If the stability of our environment is uncertain, how can we find purpose or meaning in our everyday lives? Can we still

strive for personal fulfillment while confronting the collective threat of climate change?

3. **What does justice mean in a warming world?** Who bears the moral responsibility for climate change, and how should we address the disproportionate impacts on vulnerable populations and communities?

4. **Dive deeper.** Read through each question and your responses again, and now answer the following questions:
   - Why do I believe this?

- What evidence supports my belief?

- Could this belief be oversimplified?

- What other factors should I consider when I think about this issue?

– How do I feel as I explore these topics?

____

____

____

____

____

____

____

Bayo Akómoláfé, PhD, defies labels; he is frequently described as a philosopher, poet–writer, and expansive cultural thinker. He also serves as a guest faculty member in the Climate Psychology Certification program at the California Institute of Integral Studies. He offers this recommendation for confronting climate change with more of our innate human capacities engaged, and at first his paradoxical guidance is confusing: "The times are urgent: Let's slow down" (Akómoláfé, 2025, para. 1). Born in Nigeria and steeped in Yoruba teachings, Akómoláfé also is trained in clinical psychology. He explains that slowing down is less about time and more about directing our awareness. It is about "questioning our questions" (Akómoláfé, n.d., para. 8): Instead of rushing toward answers, we need to practice deep listening. He emphasizes that solving our planetary crisis does not require a specific mental state or attitude; instead, it is a function of increasing our ability to be present. He suggests that instead of ramping up and diving in, we need to witness for ourselves the transformative potential of stillness and respond to the call of wonder. A more expansive quality of presence will allow us to develop a clarity of vision that encourages restorative and potent connections and actions to emerge naturally. This is wise advice for how we can continue discovering our blind spots and uncover deeper and more valuable ways of being and knowing.

The practice exercises in this chapter help us develop what S. Boyd (2023) called *constructive uncertainty* so we can continue looking more deeply and reflect on unexamined beliefs, helpful ones in particular, and especially in those moments when we are certain that we're certain. Think of it as adding an internal question mark to our thinking where we would normally use an exclamation point. Practice: Deep Questions in a Changing World helped us recognize our own thought patterns; Practice: Naming Your Thoughts and Feelings, presented in the next section, focuses on biases that can influence our relationships with others, and later in this chapter Practice: Deep Listening guides us to develop more intimacy with our surroundings.

SLOWING DOWN INSIDE

I'm sure that your email, like mine, contains a combination of useful information and junk. Our thoughts are no doubt the same: There is a similar mix as we go through our days, taking in and replaying conversations and relational experiences, being assaulted by both subtle and overt advertising, and receiving tremendous volumes of sensory input. Unlike with our email, though, we often don't take the time to sort through and make sense of everything our brains are receiving. The following exercise will show you how to slow down and note the quality and origins of your thoughts and become aware of the space between and around them. It is a mindfulness exercise that will help you gain perspective by keeping it simple, focusing on one thing at a time in a neutral, nonjudgmental way. You will eventually arrive at an effortless understanding of what is present in your body, mind, senses, and feelings.

Practice: Naming Your Thoughts and Feelings

Set aside 10 minutes for this practice exercise. The time can be extended once you are familiar with the practice and start to discover its benefits. Sit or recline in a quiet, comfortable setting that is free from distractions. Close your eyes, or keep them open but with a soft downward focus.

- For the first 2 minutes, place your attention on your breathing, and name the physical movements that you notice happening internally: "Breathing in, breathing out; rib cage and belly expanding, and now receding." If your mind wanders, which our minds are prone to do, simply bring your focus back to your breath.
- Now draw your attention to your thoughts and silently name them in a similar way for about 3 minutes: "Here's a thought about (planning . . . remembering what happened yesterday . . . this practice)."
- Take particular note of thoughts that contain judgments: "Here's a judging thought about (my performance . . . my body . . . my partner)."
- Now shift the focus to silently naming your emotions for 2 minutes: "This is (boredom . . . worry . . . joy)."
- Finally, notice the space after one thought ends and another begins, and note any of the qualities you perceive.
- Close your practice by releasing the naming exercise, bringing your attention back to the room, and opening your eyes.

Spend a few minutes writing about what you experienced.

1. How often were your thoughts and feelings colored by climate concerns and eco-emotions? Track the trends of your thoughts over several practice sessions and record them.

____

____

____

____

2. What were you most aware of?

____

____

____

____

____

3. Did anything surprise you?

____

____

____

____

____

Practice: Naming Your Thoughts and Feelings is not about what you should or should not think or feel; it is an opportunity to become more aware of your patterns of attention and sense more clearly the qualities of life beyond your thoughts. The practice may ultimately strengthen your convictions, shift your perspectives in surprising ways, add subtlety to your understanding, or bring stability to your emotions.

SLOWING DOWN IN RELATIONSHIPS

I am often asked by passionate but frustrated climate stewards what they can say to convince others about the urgency of the climate crisis: "Is there a better wording, tone, or timing that will make me more persuasive?" This might be a good time to take Akómoláfé's

(2025) suggestion about questioning the question. Our desire to get others involved in climate action is certainly worthwhile; however, sometimes the question embodies the problem and compromises the desired outcome. Deborah Tannen (1998), a professor of linguistics at Georgetown University, studies the mechanics of conversation and described it this way: "Smashing heads does not open minds" (p. 26). Even the right message, delivered at the wrong time or in the wrong way, becomes the wrong message.

> *Even the right message, delivered at the wrong time or in the wrong way, becomes the wrong message.*

One of the most overlooked aspects of good communication is listening. Even if we choose our words skillfully, people can usually sense a certain energy when we are trying to convince them of something: an us-versus-them, savior-versus-saved message that can feel like being scolded. It's easy to communicate that "I'm right, you're wrong, and you need to change," even when those words aren't spoken. If we feel attacked and judged, it often triggers our defensiveness and resistance. We can't invite dialogue by delivering a speech.

In addition to the content being shared, deep listening also attends to what is unsaid: the layers of meaning, the emotional tenor, the context, and the nonverbal communication. This may be uncomfortable for you, especially if you are triggered by positions with which you disagree—and many of us automatically avoid discomfort and awkwardness in our connections with others, so it might take some practice. This is a preparatory practice for a full dialogue in which you also share your own perspectives.

The following exercise slows down our conversations by demonstrating our empathy and offering our full presence, which allows the speaker to feel heard and understood. This approach can soften defenses and foster a deeper connection.

Practice: Deep Listening

This practice will take 5–10 minutes.

Ask a friend or family member who has a different view of climate change if they have 5 minutes to share their perspective with you. You can make a general request, such as "Would you be willing to share your thoughts about climate change?" or a more specific prompt: "I'd like to hear your thoughts about the upcoming vote on a clean air initiative for our community." Choose someone outside your comfort zone. Your role for 5 minutes is to listen without responding, and when the time is up thank the person for sharing with you.

Now write about your experience with deep listening.

1. What did you learn about the other person?

____

____

____

____

____

2. What did you learn about yourself?

____

____

____

____

____

3. Did you discover anything surprising?

____

____

____

____

____

Although it is important to have our voice heard, we must do so in ways that do not perpetuate or deepen the divide—and that is where the gift of slowing down comes in. Viktor Frankl was an Austrian neurologist, former Nazi concentration camp prisoner, and prolific writer whose ideas have influenced generations of thinkers. Stephen Covey, drawing on Frankl's work, captured this spirit: "Between stimulus and response there is a space. In that space is our power to choose our response. In our response lies our growth and our freedom" (Covey, 1989, p. 69).

> *Although it is important to have our voice heard, we must do so in ways that do not perpetuate or deepen the divide—and that is where the gift of slowing down comes in.*

COMMUNICATION THAT CAN BRIDGE OUR DIVIDES

When you want to introduce a new climate-related idea to someone whose opinion differs from yours, frame it as something you are offering, not asserting. Instead of saying, "Here's why you're wrong" or "The real issue is . . . ," try something like, "Can I share a different angle that helped me make sense of what's happening with the climate?" or "Would you be open to hearing something that shifted how I think about this?" This gives the other person a sense of agency and lowers the stakes: They don't have to agree; they're just listening. Whatever you share, keep it grounded in your own experience or values rather than firing facts like arrows. For example: "For me, learning about how extreme heat affects our local schools made climate change feel closer to home." Experience invites empathy; facts alone can invite debate.

You can also connect your idea to something the other person already cares about. If they value economic stability, frame climate solutions in terms of reliability, jobs, or innovation. If they value family or community, highlight safety, health, and shared flourishing. Joining their values before adding your viewpoint helps the new idea land as an extension of the conversation, not a counterargument. The key is to stay curious, stay relational, and keep the energy focused on mutual understanding rather than persuasion.

THE MOVEABLE MIDDLE

The Yale "Six Americas" model (Leiserowitz et al., 2025) presents a spectrum of climate opinions in the United States that are based on a person's beliefs, level of concern, and engagement with climate change. The model includes six groups: Alarmed, Concerned, Cautious, Disengaged, Doubtful, and Dismissive. Instead of a binary divide of believers versus deniers, the framework shows just how nuanced people's views actually are. At one end of the spectrum are Alarmed people, who are highly engaged and eager for action. At the other end, Dismissive people reject climate science and are often bound up in identity, ideology, or distrust; in many cases, acknowledging the science would challenge deeply held beliefs about who they are, where they belong, or how the world should work. The most fertile ground for meaningful conversation sits in the middle: the Concerned, Cautious, and Disengaged people. These groups are not denying climate change; they often are unsure what to believe. They may be overwhelmed, or they are simply lacking information that feels relevant to their daily lives.

This middle space, the "movable middle"—in this case, the Concerned, Cautious, and Disengaged groups—is where some of the most rewarding and effective conversations happen (Leiserowitz et al., 2025). People in these groups tend to be curious, caring, and open to learning, even if they are not yet acting on climate concerns. Research shows that tailored, empathic dialogue can boost understanding, increase personal relevance, and promote meaningful engagement in these middle groups (Goldberg et al., 2019; van der Linden et al., 2015). When climate issues are linked to values people already hold, including health, family, economic stability, and local safety, the shift from passive concern to motivated action becomes much more likely. In addition, because these groups make up a large share of the U.S. population, even small movements here can significantly shift the needle on public support for climate solutions (Leiserowitz et al., 2025). Conversations in this middle space are like planting seeds in soft soil where they will tend to take root, grow, and influence broader networks in subtle but powerful ways.

Here are a few examples of the kinds of questions you can bring into these conversations to help connect climate issues to what matters in the everyday personal lives of the people around you:

- When you think about the future, what feels most worth safeguarding for your family or community?
- What experiences have shaped the way you think about changes in the environment?
- Are there parts of the climate conversation that feel confusing, overwhelming, or unclear?
- What are some challenges you think most people in your community agree on, regardless of their politics?
- What kinds of solutions feel reasonable or worth exploring, even if you're not sure yet?
- If you could improve one thing about your environment or community, what would you choose?
- Where do you feel you have influence or a voice, even in small ways?
- Is there anything you've been curious about but haven't had the chance to dig into?
- What would feel like a helpful next step in understanding this issue?

SLOWING DOWN IN OUR ENVIRONMENT

Just as our climate conversations call us to slow our pace and inhabit a deeper presence, nature calls us in the same way. In my professional practice, I have repeatedly witnessed how this slowing opens new ways of perceiving and relating. The story I share next shows what becomes possible when our awareness meets the living world with slowed-down attention in nature.

While hosting a Climate Listening Circle, I received an unexpected and beautiful lesson from one of the participants, an insight that emerged from a deep listening to life itself. I arrived early so I could prepare a guided meditation centered on the image of a tree—a symbol of resilience, rootedness, and connection with the natural world.

Just before we began, one of the regular attendees, a passionate young activist in her 20s, burst through the door, a little breathless but glowing. She already knew the theme of the session and exclaimed, "As I was walking in, I suddenly noticed how many magnificent trees surround this building—and I'd never really seen them before! And now here I am, heading into a windowless room to close my eyes and imagine a tree!" She laughed, then grew thoughtful. "But honestly, I think something shifted. The insights from our visualizations must have been building quietly. Out there, something clicked—I slowed down and felt present, awake. I realized I could bring the same presence we cultivate in guided imagery to the real world. I stood still, truly present with the trees—calm, curious, open. No assumptions. No distractions. Just deep seeing. And in that moment, I understood: These trees, like all of life, are alive, changing, in relationship with us. Everything felt different."

She radiated aliveness as she spoke, and her realization became a gift for the whole group. We could see how this deepened awareness nourished not only her activism but also her whole being. And it reminded us all: When we slow down and attune to the living world, we not only build resilience—we remember what we are fighting for. Even in the work of systemic change, we can stay rooted in wonder. The following exercise offers an opportunity to explore this more deeply.

Practice: Being Present With a Tree

Set aside 10–30 minutes for this practice.

Choose a place outside: in your yard, a park, or any natural space where you can safely and comfortably spend time and relax. You will be finding and keeping company with a specific tree, so the first step is to locate one that you are drawn to. Spend a few moments looking or walking around; see if there is a tree that naturally draws your attention. Next, find a way to be near it, whether you lean back against the trunk or sit on the ground or a bench nearby.

Begin by taking a few full breaths, setting aside any thoughts about what happened earlier in your day and whatever awaits you later. Allow your breathing to invoke a feeling of being present in your body, head to feet, and sense your body as it exists in this particular place and in this particular moment.

Now, start to take in the whole tree: its height, shape, colors, and textures, sensing its roots and its place in the larger environment. Even if the tree is very familiar to you, can you look at it like you are seeing it for the first time? What smells do you notice? Be aware of any sounds around you and the tree, close in and then farther away. What do you hear? Birds, insects, leaves moving in a breeze, your own breath? Feel the sensation of the air on your skin and the temperature of the sunlight or shade. If you would like to feel the texture of the tree's bark, or of any moss or lichen that may be living on the tree, go ahead and do that now.

This is an exercise in sensing and feeling, so if you find that your thoughts are shifting into a story or commentary about the tree or other distractions arise, kindly let them go when you notice it happening and return to your connection with the tree.

Next, consider that being with this tree is also an energetic exchange. What qualities of aliveness do you sense emanating from the tree? How do you feel in the presence of something that has been here longer than you, and may remain long after? Take a moment to wonder: If this tree had a way of sensing you, how might it be experiencing your presence? Notice any subtle shifts in your breathing, your posture, or your mood as you settle into this relationship. Take some time with this felt sense of connection.

We know that communication is not always verbal or literal. If the tree has a message or something it wants you to know, what is it sharing with you?

When you are ready to close end this practice exercise, note any shifts in your body, mind, heart, and quality of awareness. Thank the tree and the ecosystem of which it is a part in a way that feels right to you. Take time to recognize that in the same way you can carry a glass of water from outdoors into your home, you can carry the connection and qualities you have just sensed away with you as you shift from this environment and this practice into the next part of your day. Take a few moments to write about your experience. What felt most true and meaningful? If this experience were offering you guidance, what might it be telling you?

The practice exercises in this book help us slow down, heighten our awareness, and move us out of reactive states. Practicing different forms of mindfulness can be very valuable in working with eco-emotions and facing climate change—but it is not always easy because of the well-established pattern that pulls us back to our thoughts and the fast pace of our Western culture that reinforces that pattern. Just ask my clients, who sometimes share humorous stories, like the one about getting a speeding ticket on her way to a meditation retreat, or another who was so focused on repeating the mantra "Stay in the present moment" that she passed the store she was walking to! Let's make lots of room for ourselves and others, holding our life lightly and offering kindness as we stumble toward new ways of being.

In Chapter 7, we will learn why finding and building community is essential when we are navigating the emotional and practical realities of a changing climate. None of us are meant to do this work alone: Community offers shared resilience, collective wisdom, and the grounding reminder that we belong to something larger than our individual fears or efforts. We also will explore clear, doable guidelines for cultivating these connections and working with the predictable challenges that arise in groups, whether you are joining a local climate initiative, forming a circle of mutual support, participating in neighborhood resilience projects, or simply strengthening trusted relationships that help you stay balanced. Community isn't something that is nice to have in our rapidly changing climate; it is one of the most powerful tools we have for sustaining our courage and creativity and taking meaningful action.

CHAPTER 7

FINDING AND BUILDING COMMUNITY

For much of its history, mental health treatment has focused on improving our individual well-being by addressing personal struggles and trauma and building up our coping mechanisms. However, climate change is affecting us all on a broad scale, and the resulting eco-distress is prompting a shift in how we seek relief. Mental health and resiliency practices are moving into political and social arenas in ways we have not seen before. Collective trauma calls for a collective effort to heal, and that includes a more community-centered ethic of mental health and resiliency as well as conversations about the sources of environmental destruction and the policies necessary to remedy it (Cosentino et al., 2025). In this chapter, we consider the benefits of finding a community that shares your desire to address climate change. We explore ways to find a community and ways to navigate the inevitable differences that arise within the community.

JON'S STORY: THE OVERLOOKED NEED FOR COMMUNITY

Being part of a community can be deeply fulfilling, but this can be an overlooked need in our hyperindividualistic Western culture. Consider the story of one of my therapy clients, Jon: His unrecognized longing to connect with others around his climate distress led to unexpected insights that emerged in a therapy session that was initially focused on his chronic back pain.[1]

"Hmm. . . . Where should I sit?" Seating arrangements are the first dance in therapy, and Jon is eager to know the steps. He is a sensitive bundle of restlessness and exhaustion: His feet tap anxiously after he sinks into the couch, but his eyes look exhausted and ready to close. Jon is a first-generation Chinese American in his early 30s and a new graduate with dual degrees in geology and biology. He mentioned having tried "therapy lite" at a university health clinic. Since starting his first job, a persistent backache has puzzled his doctors, and they directed him to my office. He also wanted to work with a climate-aware therapist: He is a dendrochronologist whose work involves measuring and recording tree rings to determine a tree's age and understand past environmental conditions.

Trees are scribes. Many of us remember learning as a child that you can count the rings of a tree to learn its age. Jon was eager to describe how they also record a reliable history of the region's climate conditions over long periods of time. Around 1960, tree rings around the world began manifesting new patterns with abrupt shifts between narrow and wide bands that reflected more and more extreme weather events. This kind of planetary shift had never been seen in the 290 years that scientists had been studying tree rings (Trouet et al., 2018). Jon has been watching both climate history and our future unfolding before his eyes every day in an ancient organic language that he is an expert at interpreting.

As Jon describes the details of his work stress, it becomes clear that his devotion to the environment brought him to his job, but he is feeling restricted by his organization's sluggish bureaucratic pace and cumbersome procedures. Every day, he records the story the tree

---

[1]Jon is based on a real client, but his name and some identifying characteristics have been changed to protect confidentiality.

rings are telling him, and his body tightens as he explains how those urgent data seem to disappear into a sort of black hole: Warnings go unacknowledged, unacted upon, and, in effect, are lost. As he speaks to me, his voice rises, and his agitation grows; it is as if he has climbed onto a soapbox with a megaphone, scanning the crowd for anyone who might care that the trees are sounding an alarm about the direction humanity is headed. As his voice sounds the alarm, I wonder if its shrilling is also reverberating down his spine and into his back. In somatic therapy, we pay close attention to these kinds of signals, the way a story lives not just in words but in the body, because they often reveal deeper layers of activation, stress, or unprocessed emotion that are seeking acknowledgment (Ogden et al., 2006).

After a few sessions, I invite Jon to explore his persistent back pain using nonscripted guided imagery. The process resembles hypnosis, but rather than a scripted scenario designed to lead to a particular outcome it is an open and direct exploration of a person's thoughts, feelings, and body sensations. The frontal lobes of the brain's cortex are used for advanced cognitive activities, such as planning, and in nonscripted guided imagery that part of the brain is quieted. Activation shifts to the posterior cortical "hot zone," a region that integrates sensory information and supports specialized forms of data processing. Symbols and images are the natural language of this part of the brain, which is also active when we dream (Siclari et al., 2017). The posterior cortical network is invaluable, not because it is more important than our ability to reason but because we have stopped using our whole brain, including the areas that offer insights we cannot access through rational thinking alone.

When we experience pain, we usually scramble to distance ourselves from it, but the guided-imagery process does the opposite. The client is guided to become more intimately attuned to their sensations, and this often results in discovering keys that will unlock a deeper, more lasting resolution.

I invite Jon to turn his focus inward: "Close your eyes and take a few breaths, gradually allowing your breathing to slow and deepen. Welcome the fresh oxygen into your body as you inhale, and find a letting-go quality as you exhale to help your body relax." As Jon's body, mind, and emotions settle, I ask him to bring his attention to the sensations in his back: "Trace the edges of your pain, using your awareness like a pen. Draw it in three dimensions in your mind's eye. Keeping your eyes closed, describe the shape aloud."

Perhaps it is no surprise that Jon reports that his pain appears to be "like a tree" whose trunk runs up his spine and whose branches extend under his left shoulder blade. But the strongest sensation felt like "roots twisted tight" across his lower back. After exploring the colors, textures, size, and other features of this tree, I ask Jon to give it a voice, some way for the tree-pain to express itself to him. We start with the most intense sensations: "What do the twisted roots want you to know?"

Jon begins: "They're telling me that I'm missing something in my work: that understanding the rings is important, but there's more to it. The roots want me to pay attention; they want to show me something."

I listen, allowing plenty of time for him to stay with his unfolding process. I encourage him to continue exploring the messages that are emerging. His face softens, and suddenly he looks very young.

Jon begins describing the *mycorrhizal network*, the complex symbiotic relationship that trees have with microorganisms in the soil. His insider references to fungi networks and bacteria go over my head, but I can see that he is clearly feeling invigorated by reconnecting to this knowledge, which is part of his training but not part of his job. "Trees communicate with each other and share nutrients through their roots," he explains, and goes on to describe how trees experience *kin recognition*—recognizing and favoring related trees—through the mycorrhizal network and how this relationship increases the trees' health and resilience.

When this imagery exercise is complete, Jon is quiet for several minutes. When he opens his eyes, the flashing pain has converted to flashes of insight. For the moment, his back is not as tight, and the pain has subsided.

Jon fills me in on what happened during that silence. He realized that he is isolated, like "a tree growing in a pot instead of a forest." He has not formed networks with other "kin" who share his passion and vision and with whom he can share symbiotic ideas and feelings. He is now determined to find groups of scientists and activists whom he will choose to join as an important step in his personal growth. He will begin feeling out new ways to spread his roots.

Practice: Reflection and Connecting Within

1. Does Jon's story bring up any particular feelings for you?

____

____

____

____

____

2. Even if your work or life circumstances are completely different, are there elements of his story you can personally relate to?

____

____

____

____

____

3. Jon's insight emerged as he quietly tuned into the sensations in his body. Close your eyes, take a breath, and spend a few moments tuning into your own. Do any images or impressions come up? If so, what do they want you to know?

____

____

____

____

____

COREGULATION: OUR NERVOUS SYSTEMS TALK TO EACH OTHER

Did you know that emotions and emotional states, such as calm and distress, are contagious? This idea is supported by a theoretical model in the field of psychology called *polyvagal theory*, which was developed by Stephen Porges (2011), PhD, Distinguished University Scientist at Indiana University and founding director of the Traumatic Stress Research Consortium. A key concept in this theory is *coregulation*: the idea that our nervous systems are biologically wired to attune to one another (Porges, 2011; Porges & Dana, 2018), similar to the roots of trees in mycorrhizal networks. Through cues like tone of voice, facial expressions, eye contact, and even breathing rhythms, our bodies read the emotional states of others. When someone is calm and regulated, their nervous system can help guide another person's nervous system toward greater calm and safety.

This is like an invisible dance that happens largely outside our conscious awareness in which our nervous systems are constantly adjusting in response to the people around us. Coregulation is especially important because it helps us manage stress and maintain emotional balance. We cannot fully regulate ourselves in isolation; we need others, especially during times of distress. Like many of the ideas we have been exploring, polyvagal theory affirms how deeply interconnected we are: As we strengthen our own resilience, those around us benefit, and in the presence of others we can more easily expand our own capacity to regulate. In this respect, self-care and community care are one and the same.

FINDING COMMUNITY

Creating or joining a community can offer a wide range of benefits. It can foster a sense of belonging and solidarity and provide both emotional and practical support. Depending on the nature of the group, it can also give us opportunities to pool resources and solve problems together, while adding an element of safety through mutual care. Research shows that social connection is strongly linked to better mental health overall and reduces the risk of anxiety, depression, and stress (Holt-Lunstad et al., 2010). Being part of a supportive group helps us feel heard, understood, and valued, which boosts our self-esteem and emotional well-being. Climate action communities, for example, can be a powerful force for change. A community also offers a space for shared learning and growth where we can find encouragement to stay focused on and accountable to our goals. By joining forces, individuals amplify their voices and have a greater impact than when acting alone.

> *By joining forces, individuals amplify their voices and have a greater impact than when acting alone.*

There are many different types of communities that can offer support and connection. It can be as simple as regular get-togethers with a few like-minded friends or colleagues where you can be yourself, share openly, and listen to each other's questions and feelings. A community can also be defined by geography; for example, groups may form in your neighborhood or region to work on local environmental initiatives. Some may revolve around shared culture, language, traditions, or heritage. Faith-based groups that focus on environmental stewardship also offer a strong sense of belief-driven purpose. When you feel drawn to engage, look for communities that prioritize trust; emotional intelligence; a shared vision; collaboration; sustained engagement; and a commitment to diversity, justice, and inclusive participation.

NAVIGATING DIFFERENCES WITHIN A COMMUNITY

Although the concept of a community sounds ideal, and they are often welcoming and friendly, they come with their own set of challenges. This is especially true in diverse groups whose members come from different backgrounds with very different experiences, identities, and worldviews. For example, some cultures value direct communication, whereas others prefer indirect or more subtle ways of expressing thoughts and feelings. These differences can sometimes result in misunderstandings or misinterpretations of intent. Building a strong community takes continuous effort, reflection, and a commitment to nurturing meaningful relationships.

Several leaders have offered powerful guidance on how to navigate our differences generatively. Among them is Resmaa Menakem (2017), PhD (*honoris causa*), author of *My Grandmother's Hands: Racialized Trauma and the Pathway to Mending Our Hearts and Bodies*. He reminds us that working across difference is not just an intellectual task: It also requires tending to the somatic residue of trauma held in our bodies. Menakem distinguishes between *clean pain*, the discomfort we feel when we confront hard truths, take responsibility, and stretch beyond what is familiar, and *dirty pain*, which arises when we avoid that growth through blame, denial, or defensiveness. Building the capacity to collaborate across differences means continuing to expand our nervous system's window of tolerance—the zone in which we can stay regulated and engaged—so we can remain present with discomfort instead of shutting down or acting out (see Chapter 2, this volume).

Author adrienne maree brown (2017, 2020), who wrote *Emergent Strategy: Shaping Change, Changing Worlds* and *We Will Not Cancel Us: And Other Dreams of Transformative Justice*, added that meaningful change begins not at the level of nations or algorithms but rather in our relationships, in how we show up for each other in joy, in tension, and especially in conflict. "How we are at the small scale is how we are at the large scale" (brown, 2017, p. 52) she wrote, drawing on the idea of fractals to show that the patterns we practice in the micro reflect and shape the macro. Echoing Bayo Akómoláfé's (2025) call to move slowly (see Chapter 6, this volume), brown emphasized that collaboration can't be forced or rushed. It unfolds "at the speed of trust" (brown, 2017, p. 41) through listening, consent, and the cocreation of safety.

The deeper roots of the social inequities we have discussed can also show up in groups, with subtle power dynamics causing hierarchies to form. In diverse groups, people with higher levels of privilege or power—based on their race, gender, experience, or job title—may unintentionally dominate conversations or decision-making, leaving the voices of people whose expertise does not carry institutional status unheard. To address this issue, Vanessa Machado de Oliveira (2021) offered a helpful framework for collective action she calls stepping forward/aside. Machado de Oliveira is the dean of the Faculty of Education at the University of Victoria in British Columbia, Canada, and the author of *Hospicing Modernity: Facing Humanity's Wrongs and the Implications for Social Activism*. She defines *stepping forward* as actively engaging, taking responsibility, and participating in efforts to create change. On the other hand, *stepping aside* involves recognizing when to pull back, reflect, or listen in a way that opens space for others to lead and shape the way forward, especially those whose voices have been historically silenced. She reminds us that the comforts, securities, and pleasures of more privileged groups continue to be sustained by the expropriation and exploitation of marginalized people, species, and land.

The next practice exercise will guide you to explore ways to be thoughtful, responsible, proactive, and self-aware in community settings and build a skillful attunement to diversity. Like all the perspectives and practice exercises in this book, this one is an opportunity to approach community and eco-action with humility, discernment, and self-awareness. You will explore *positionality*, which involves recognizing how aspects of your social identity, such as your race, gender, class, age, ethnicity, ability, sexuality, and other factors shape your perspective, experiences, and interactions with the world. The concept of positionality acknowledges that everyone's worldview is influenced by the social positions they occupy, with each carrying different levels of privilege or disadvantage. Understanding and acknowledging your positionality helps you become more aware of the biases that may inform your feelings and actions. In a community setting, this awareness helps us foster trust and connection.

Practice: Exploring Your Positionality

As you respond to each question, include a few ways that you can step forward or step aside according to Machado de Oliveira's (2021) guidelines.

1. Name your position, listing your race, gender, class, age, ethnicity, ability, sexual orientation, and other factors that

shape your perspective, experiences, and interactions with the world.

2. Are you consciously aware of and accepting of your position? Why or why not?

3. What systems, institutions, or social structures either work smoothly or do not work well for you—either directly or indirectly—because of your privilege or lack of it?

4. In what spaces or situations—school, work, community gatherings—do you feel at ease or uncomfortable, and what might be contributing to that comfort or distress?

5. How does your privilege, or lack of it, shape your relationships with people in other ethnic, racial, age, and socioeconomic groups? Do you notice any power dynamics at play?

____

____

____

____

____

6. What actions can you take to acknowledge your position in ways that are supportive of social justice and equity?

____

____

____

____

____

7. How do you feel when your position is unrecognized, questioned, or challenged? What emotions come up, and why do you think that is?

____

____

____

____

____

Although creating a truly equitable community can be challenging, it is far from impossible. The basic ingredients—well-defined goals, clear communication, inclusivity, empathy, continued learning, and trust—can help overcome many obstacles. If people feel they can rely on one another, they are more likely to fully engage and contribute. Building trust is a gradual process that entails community members demonstrating reliability, openness, and honesty over time. When these ingredients come together, the result is a vibrant and resilient community where members feel connected, supported, and empowered to work toward a shared goal.

> *If people feel they can rely on one another, they are more likely to fully engage and contribute.*

Practice: Appreciating Community

1. If you are already involved in a group or community, could any of the principles discussed in this chapter be useful for your group?

____

____

____

____

____

2. What benefits do you derive from your community, and what are you grateful for?

____

____

____

____

____

3. If you are thinking about joining or starting a community, what attracts you to one type of group over others? Do you feel any hesitation and, if so, about what? What is one small step you can take to begin exploring the group?

____

____

____

____

____

MULTISOLVING

Let's look at some ways for you and your community to bring eco-justice practices into your climate work. One of the most promising frameworks for climate-related systems change is Elizabeth Sawin's (2024) model of *multisolving*, which emerged from her background in biology and systems thinking. Sawin, PhD, formerly cofounded the think tank Climate Interactive (https://www.climateinteractive.org/) and is the author of *Multisolving: Creating Systems Change in a Fractured World.* You have read about how climate change comprises many layers of complex challenges that are intertwined to form a global "wicked problem" (Rittel & Webber, 1973). Sawin's approach focuses on creating coordinated plans that address the various tangled issues simultaneously instead of individually. One example of this kind of holistic and integrated solution is the implementation of community-led clean-energy projects in underserved areas.

Access to reliable and affordable energy can be a major challenge in low-income communities. By investing in locally controlled renewable-energy projects—often financed through public funding, nonprofit partnerships, philanthropic grants, or community-ownership models rather than out-of-pocket individual contributions—communities can build solar farms, microgrids, and wind projects that do not rely on wealthier residents to foot the bill. These initiatives decrease dependence on fossil fuels and reduce the carbon emissions that are driving climate change. They create cleaner air, local job opportunities, and more affordable energy while also increasing energy resilience during the extreme weather events that are becoming more common. The Cooperative Energy Futures solar garden at Shiloh Temple International Ministries in north Minneapolis, Minnesota, is a good example: It offers credits to low-income households on their energy bills and uses local laborers who have been trained in community job-development programs (Low-Income Solar Policy Guide, 2023). Projects like this show how community-run clean-energy systems can cut power costs, strengthen local infrastructure, and advance both climate solutions and social equity.

Embedded in this example are some key principles. First and foremost is the basic precept "No decision about us without us" (Sawin, 2024, p. 201). This principle is vividly expressed in the Northern Cheyenne Tribe's White River Community Solar Project in Montana. The tribe secured funding from the U.S. Department of Energy; retained ownership of the systems; and incorporated tribal governance, workforce training for community members, and decision-making authority over who receives solar installations and how benefits are distributed (Indigenized Energy Initiative, 2022).

Sawin (2024) also offered other strategies to discuss in collective planning conversations:

- "Who benefits, and who's burdened?" (p. 200)
- How and where can we access the power to change how resources are distributed?
- How can we disrupt the self-reinforcing feedback loops that deliver ongoing "success to the successful?" (p. 201)

This systemic approach is a powerful example of how we can integrate our internal resources—our thoughts, emotions, and nervous systems—with external resources, such as climate action and community engagement, to face and heal our climate-related emotions together. Collective healing involves recognizing the profound pain; honoring the community's vision and resilience; and working with others to create a future that is grounded in mutual understanding, healing, and empowerment. Communities can push for systemic change that transcends individual lifestyle choices to foster a collective reckoning and rebuild trust, both within the community and with other groups.

In Chapter 8, we take a deep dive into a different kind of connection and sense of belonging: our kinship with nature. The practice exercises in Chapter 8 are designed to help you experience the reciprocal and interconnected flow of life in nature and to sense its inherent balance, which often extends beyond what logic can explain. You will discover that we are not separate from nature; instead, we are part of a vast, interdependent web of life.

CHAPTER 8

ENTERING KINSHIP WITH NATURE

Perhaps our most damaging collective myth is that we are separate and different from nature. We often make plans to spend time in nature, and we may enjoy it, fear it, crave it, ignore it, or romanticize it—but, despite our attitudes and beliefs, both human and nonhuman life flourish in complex, dynamic, finite, and interwoven systems in which we are indivisible from, and held in a delicate balance with, all of existence. Even though physics, biology, and common sense tell us otherwise, this myth of separation nevertheless fuels thoughts and behaviors that fly in the face of the natural laws that govern these systems. We fish as though the oceans were an infinite source of food supplies rather than a carefully balanced ecosystem; we burn fuel as though spewing pollution into the air does not affect our breathing; we take our garbage bags to the bin as though they'll disappear into the Great Somewhere, never to be seen again; and we lay waste to forests as though trees were simply economic commodities rather than centuries-old engineers of the biosphere that turn carbon dioxide (CO_2) into oxygen, help cool the planet, and remove pollutants from the air. Human nature and the very composition of our bodies are of the same ancient and evolving elements as the oceans, trees, and fields. Pull a thread anywhere in our biosphere and you will find that it is attached to everything else, including our thoughts and feelings.

Climate change is one of the devastating consequences of our distorted perceptions that view our living systems as inanimate and treat intimately connected processes as separate and self-sufficient. As we dismantle this myth, we uncover a natural way of seeing that allows us to participate in a restorative way of living that our species has known all along. We find a place of belonging with the land, each other, and the more-than-human world.

In this chapter, we reshape our lived experience with the natural world by exploring a kinship-based perspective inspired by, and offered with deep respect for, the teachings shared by many Indigenous cultures. As we practice attuning to the Earth, we will strengthen our capacity to move through climate challenges with clarity, compassion, and a renewed sense of belonging.

DEFINING NATURE

Nature can be difficult to define. The meaning has gone through many changes over time, and it is viewed very differently in cultures around the world. How you think and feel about nature is influenced by where and how you grew up and the sum of your life experiences. The following exercise is not aiming for a "correct" answer; instead, it is an opportunity to discover how you relate to the environment as we evolve together toward a more eco-wise relationship with all of life. Writing down our thoughts and feelings can help bring clarity, insight, and perspective to our beliefs.

As you write about nature, it may be helpful to recall some of your experiences in detail. It is possible that the way the questions are phrased won't precisely align with your thoughts, feelings, and experiences, so consider them to be very flexible prompts for doing your own exploration. Your answers will likely shift over time, and it can be fun and insightful to revisit your notes 6 months or a year from now and write about nature again.

Practice: Exploring Your Beliefs About Nature

1. How do you define nature?

____

____

____

____

____

2. Is there anything you see as separate from nature? Where does nature stop?

____

____

____

____

____

3. What feelings arise when you think about nature? Does that change when you're "in" nature?

____

____

____

____

____

4. Does nature ask anything of you? Do you ask anything of nature?

5. If you could identify three core qualities that describe the essence of nature, what would they be?

6. If nature had a voice or some other way of communicating with you, what would it want you to know right now?

These questions ask you to look deeply within and around you. After you complete this writing practice, take a few moments to notice how you feel, and write about any shifts in your body, mind, emotions, or relationship to your surroundings.

DEFINING KINSHIP

Much of what we are exploring in this book focuses on relationships and connections that expand in ever-widening circles: how we attend to our own thoughts and feelings, how we care for our communities, and how we meet the world around us. To take our relationships a little deeper, it is helpful to look at the many meanings of *kinship*. The term "kin" often points to blood relatives, but kinship can also describe an evolving web of social and emotional bonds that shape our identities, behaviors, and sense of belonging. In many Indigenous societies, kinship is understood far more expansively than in Western cultures: Everyone within the clan or community is regarded as a relative to whom we are linked through mutual responsibility, reciprocity, and lifelong support. Drawing from both Potawatomi and Anishinaabe worldviews, Robin Wall Kimmerer (2013) described how this understanding of family extends beyond humans to include the land, waters, plants, and animals with whom people have cultivated relationships over tens of thousands of years. There is so much we can learn from these deeply rooted perspectives when we come to them with openness, humility, and respect for the communities that carry this knowledge forward.

A scientist, member of the Citizen Potawatomi Nation, and founder of the Center for Native Peoples and the Environment, Kimmerer (2013) deepened this teaching in her book *Braiding Sweetgrass: Indigenous Wisdom, Scientific Knowledge, and the Teachings of Plants*, in which she explained that "ki" (singular) and "kin" (plural) are variants of the Anishinaabe word "*aki*," meaning "earthly being" (p. 301). She invited us to reclaim a language that reflects our interdependence with the living world, calling this the "grammar of animacy" (p. 55)—a way of speaking that honors the vitality and agency present in all beings. This kinship view cultivates a sense of responsibility, reverence, and care: an understanding that the well-being of humans is inseparable from the well-being of the earth. Such a perspective offers guidance for building more sustainable societies and embracing an ethic of environmental stewardship that is relational rather than extractive.

Here are just a few of many examples from Indigenous cultures around the world:

- One of the best-known examples of kinship with nature comes from the Haudenosaunee (Iroquois) culture. The Three Sisters—corn, beans, and squash—are regarded as family members working together in harmony, and their growth is considered a gift that requires our responsibility to tend to them (Mt. Pleasant, 2006).
- The Lakota, a group of Native American peoples from the Great Plains, have a worldview based on the sacred circle in which all beings—human, animal, plant, and elemental forces—are considered equal members of a family (Marshall, 2001).
- The Māori people of New Zealand have a strong sense of kinship with the natural world that is reflected in their concept of *whakapapa* (genealogy). Whakapapa connects humans, plants, animals, and the environment within a vast interconnected family. The Māori have long recognized that humans are descended from the natural world and that all living beings have ancestors and relationships that tie them together (Roberts et al., 2004).

These kinship connections are often reflected in the local language. The Inuit (in Arctic regions of Siberia, Canada, Greenland, and Alaska) refer to the stars as "brothers" and "sisters," expressing a relationship that acknowledges their role as guides and helpers (MacDonald, 1998). In many African traditions, rivers are regarded as life-givers and spiritual beings. A river may be called a "Sister River," symbolizing a nourishing, familial connection with the people it sustains (Gyekye, 1995). In Taoism and other East Asian spiritual practices, the sun and the moon are often viewed as parental figures and are referred to as "Father Sun" and "Mother Moon" (Pregadio, 2008).

As we look across cultures and the millennia, kinship teachings express the many ways humans have understood the more-than-human world as part of an extended family. Engaging with these perspectives can be deeply enriching and offer ways of seeing that expand our sense of belonging and remind us of our responsibilities within the web of life. But when we approach these traditions, we have a responsibility to do so with care and discernment about the difference between genuine respect and unintended harm. *Cultural appropriation*—using or borrowing cultural practices, symbols, or knowledge from marginalized communities without understanding their context or honoring their origins—can occur even with good intentions. That is why it is essential to learn from these worldviews with humility and a commitment to honoring the communities and traditions where they originate. In an interview in *Biohabitats*, Kimmerer clarified the issue:

> Someday, I would like to see indigenous knowledge and environmental philosophy be part of every environmental curriculum, as an inspiration to imagine relationships with place that are based on respect, responsibility, and reciprocity. . . . There is a tendency among some elements of Western culture to appropriate indigenous culture. They have this idea that TEK (Traditional Ecological Knowledge) and indigenous ways of knowing are going to change everything and save the world. There's certainly a lot of potential. But what is most important to me is not so much cultural "borrowing" from indigenous people, but using indigenous relationship[s] to . . . catalyze

> the development of authentic relationships between settler/immigrant society and place. Not to copy or borrow from indigenous people, but to be inspired to generate an authentic relationship to place, a feeling of being indigenous to place. (Nelson, 2012, paras. 29, 47)

Let's explore how we can cultivate an even deeper connection to nature with inspiration from an Indigenous worldview of kinship. The following exercise can help you experience the environment as living, breathing relatives. As Kimmerer (2013) suggested, this is an opportunity to generate your own authentic relationship to place.

Practice: Experiencing Kinship

Find a quiet natural space—your yard, a park, a forest, or by a river—and simply sit in silence for a while. Take a few full breaths, and invite your body to relax. Close your eyes, and listen deeply to the sounds around you. Focus on the rustling of leaves, the songs of birds, or the flow of water. Allow your thoughts to soften and become present. Now, open your eyes and take in the colors and textures, the movements, the scents, and the temperature of your spot.

1. Using an expansive interpretation—allowing kinship to extend beyond people to include the land, water, air, and living beings around you—identify your kin by focusing on the natural elements around you.

____

____

____

____

____

2. Briefly describe your connection with each element and the unique gifts they provide you.

3. What are some of the ways you depend on your kin?

4. Does anything exist outside your community of kin?

5. How do your kin relationships support your health and resilience in your climate-aware life and work?

6. How can you contribute to the health and resilience of your kin relationships?

7. Reflect on and feel how you are surrounded by your more-than-human kin. What qualities do you experience as you regard your surroundings in a new way?

You might want to take a small item home with you—a pebble or twig—to help you stay connected to your experience. You could add them to the Remembrance Space you created in Chapter 4. Quietly sense within whether it feels right to bring the item with you. If it does, is there a way you want to express gratitude to the place for this gift?

THE POWER OF ART

For many of us, these concepts of kinship may feel unfamiliar because our dominant Western culture often prioritizes logic, efficiency, self-reliance, and structured problem-solving as standards of success. Although these skills are essential, they are not enough: What we truly need is the full engagement of all our capacities. To confront the complex challenges of our climate crisis, we need to engage our creativity, intuition, and holistic thinking. These capacities are often undervalued and underused, yet they can help us think outside the box, see the big picture, and bring emotional depth to our solutions.

> *To confront the complex challenges of our climate crisis, we need to engage our creativity, intuition, and holistic thinking.*

Art and storytelling are powerful ways to blend logic, imagination, and emotion. By creating art and writing, we can express feelings that are hard to put into words. When we transfer our emotions onto paper or canvas, we allow ourselves to process and release difficult emotions, like sadness, anger, or fear, bringing a sense of relief and catharsis. Creative works, such as poetry and paintings, often communicate more effectively than raw data or practical information because they provide context and reveal how information impacts real people, situations, and outcomes, making the issue more relatable. We are naturally drawn to images and stories, especially when they resonate with our own experiences.

Creative expression has a long history of profound impacts on environmental justice movements, with countless examples that have inspired awareness, driven collective action, and led to lasting cultural change. From the Resilience Murals in New Orleans (https://iamneworleansvoices.com/murals/) and Joni Mitchell's (1970) song "Big Yellow Taxi" to Richard Powers's (2018) novel

The Overstory and even Dr. Seuss's (Geisel, 1971) *The Lorax*, works of art have played key roles in bringing complex issues to the forefront and often amplifying marginalized voices. However, some of the most powerful creations come not from well-known artists but from the heart of communities themselves. Let's dive into the story of the "Accidental Artist," Eduardo Jiménez (Brigantine Living, 2025).

Raised in San Juan, Puerto Rico, Jiménez has always felt a strong bond with the ocean and its shores (Brigantine Living, 2025). He now lives in Brigantine, a small island city located just off the coast of Atlantic City, New Jersey, and every day, 40 minutes before sunrise, he sets out on 4-mile (6.5-km) walk along the beach. He began his daily walks during the COVID-19 pandemic, when going to the gym was no longer an option. As he walked, he noticed the large amount of litter along the coastline, which sparked his awareness of its harmful impact on wildlife, both on the beach and in the ocean. He learned that when ingested, marine debris can puncture animals' stomachs or create a false sense of fullness, leading to illness or starvation. Species like sea turtles and whales can become entangled in plastic waste, which can be deadly.

In response, Jiménez began collecting and upcycling trash like plastic bags, cigarette butts, and Mylar balloons into sculptures and art installations that draw attention to the urgent issue of marine pollution (Brigantine Living, 2025). Each piece is designed to inspire awareness and prompt conversations about coastal conservation while encouraging viewers to reconsider specific actions, like tossing cigarette butts or releasing balloons into the sky. His work has gained considerable attention, including exhibitions like "Seaside Salvage" at the Camden FireWorks gallery. Jiménez humbly insists that he is simply doing his part for conservation, but it is also evident that his beach walks and art are therapeutic for him. When reflecting on his morning walks, he exclaims, "I'm not gonna clean the whole planet's trash, but at least I'm doing my little part, try to help and maybe others will also do their little part to help" (Fallon & Greenberg, 2024, para. 24).

Professional artists also have a powerful role to play in driving awareness and change, as demonstrated by THE HERDS, an international, multidisciplinary, and multicultural project (https://www.theherds.org/). THE HERDS features a collection of life-size puppets that in 2025 traveled a 20,000-km route from the Congo Basin to the Arctic Circle. Through live performances, the project vividly portrays wildlife fleeing from life-threatening climate impacts, and it draws parallels to the growing number of human communities around the world that have been displaced by the climate crisis.

The puppets are created by Ukwanda Puppets and Designs Art Collective in Cape Town, South Africa, led by Artistic Director Amir Nizar Zuabi (https://www.theherds.org/behind-the-scenes). They are crafted primarily from recycled materials sourced locally in the regions along their journey. Local people along the route are trained to build and animate the puppets, and they participate directly in the large-scale performances. Audiences can attend free public performances and workshops that are designed in collaboration with world-renowned dance companies, theater-makers, and artists. Free educational resources, including open-source puppet design kits, are made available to schools, families, and communities worldwide, encouraging broader engagement.

The inspiration for THE HERDS began with Zuabi's childhood memories of watching bird migrations between Africa and Europe (Zuabi, 2024). Reflecting on the shrinking flocks and the parallel vulnerability of human migration, Zuabi envisioned a storytelling journey that would explore the deep connections between climate change and migration (Goodgoodgood, 2025). "Animals are nature's great early warning system" (5:23), he affirmed in a TED Talk, in which he declared his intention to "disturb our indifference" (5:41) and demand action (Zuabi, 2024). As Zuabi put it, "We are just a drop of water on a stone—but over time, the water shapes the stone" (Aziz, 2024, para. 16).

In the face of the climate crisis, envisioning a transformed future—a world where ecosystems flourish and communities are united and resilient—is a crucial part of realizing change. This vision dares us to look beyond current limitations, imagining a better future and taking action even when the path ahead is dark or challenging. Practicing this kind of vision can be like an initiation, igniting an inner spark that helps us grow into our full humanity and become stewards of the world we deeply care about.

CREATIVITY BELONGS TO ALL OF US

The following exercise draws inspiration from Haiku, a traditional Japanese poetic form rooted in centuries of cultural and contemplative experience. Haiku is far more than a simple structure: It is a disciplined way of attending to the world and capturing fleeting moments with clarity, restraint, and reverence. Although we won't be practicing formal Haiku, we can learn from its spirit. Its brief, a Haiku's mindfully chosen lines invite us to slow down, notice subtle details, and attune to the natural world with fresh eyes.

A classic Haiku follows a distinct structure—three lines, with five syllables in the first, seven in the second, and five in the third—as shown in these examples:

> Green leaves kiss the sky,
> Rivers hum with life below,
> A timeless story.

> Voices rise as one,
> Equality's light shines bright,
> Justice flows like streams.

Earth's wounds heal with care,
All voices rise for her health,
Roots of change take hold.

Practice: Haikus to Envision a Thriving Earth

Write a Haiku (or more than one) that describes your vision of the healthier and more just world toward which we are working. Along with the other expressive practices in this book, your Haiku-inspired writing is less about getting the form "right" and more about being present to the experience. Although traditional Haiku follows a disciplined structure, what we are doing in this Practice Exercise is using its spirit of observation and simplicity to spark creativity—aiming not for mastery but for mindful attention. The objective is to express what is inside, however it comes out. Claim your freedom to explore, feel, and create in the moment. The beauty is in the process itself, not the end result. Give it a try now.

My future vision for a thriving world:

Take a moment to reflect on the writing process and on Haiku. Without the constraints of the Haiku structure, write down the details of the future you envisioned. What stood out to you? What aspects of the world, people, and community were most inspiring?

Writing a Haiku can be a complete practice in itself, or it can be added to any other experiential activity to serve as a meaningful touchstone for the experience. Your Haiku can also be placed in your Remembrance Space.

Guerrilla art is a form of unconventional public art that is created and displayed in unexpected locations. It is often temporary, and it aims to engage people directly in surprising ways. Consider anonymously sharing your Haiku in public places as a type of guerrilla art. You might write one on a café napkin that you leave on a table, return a library book with your Haiku written on a paper bookmark, or place a Haiku sticky note on a bus bench. You will be like the guerrilla gardeners who cast seeds into vacant lots and along sidewalks to transform these spaces into green, vibrant areas that beautify their communities.

Practice: Guerilla Haiku

Write down some ideas for where you could leave a Haiku in an unexpected place. How does it feel to be a secret agent for inspiration and transformation?

As we close this chapter on kinship and creative practice, it is worth noticing how these reflections can prepare us for what comes next. Learning to slow down, pay attention, and see ourselves as part of a wider community of life cultivates the very qualities we need if we are to make meaningful connections with one another. These practices of imagination, humility, and relational awareness lay the foundation for building bridges across generations. They help us meet each other with openness and curiosity as we join collective movements for social justice and climate action, carrying our creative insights into the shared work ahead.

CHAPTER 9

CONNECTING GENERATIONS FOR ECO-JUSTICE

In my practice, I hear climate concerns from people of all ages and backgrounds, from workshop participants and therapy clients to friends and neighbors. They echo the ones I am sharing here:

- An 8-year-old girl told me, "I'm afraid I'll never get to grow up because of the way things are going on our planet."
- A 16-year-old boy said, "Sometimes I can't fall asleep because I'm afraid we're going to melt in our own home. I don't want to finish high school: What's the point?"
- A 25-year-old woman shared, "I long to be a parent and raise a family, but I dread the horrors my children would face in their lifetime."
- A 37-year-old mother confided, "Every day I feel like I'm failing my kids, because I can't protect them from all the changes that are happening so fast."
- A 51-year-old man, who had lost his home in a wildfire, wept as he told me, "One minute we had a home, memories, and a sense of safety. The next minute that was all gone. I feel like I've lost more than just a house: I've lost a part of myself, and I don't know how to rebuild from something so devastating."

- A 67-year-old grandmother asked, "What have we done? How can I let young people know how deeply sorry I am for the world that's been given to them? I want to be ready for the conversation without coming undone by guilt and sadness."

Perhaps what some of these people have expressed resonates in your own thoughts and emotions. No matter where you fall on the generational continuum, if you make the effort to understand the fear and pain that other generations are experiencing it offers an opportunity to bridge the generational divide on the issue of climate change. This book combines emotional processing with learning how to take actionable steps—so let's continue this dual approach and address the generational fractures that climate change creates.

WHY YOUNG PEOPLE HAVE THE MOST CLIMATE ANXIETY

Everyone is susceptible to climate-related distress, but certain age groups are more strongly affected. Recent studies have shown that Generation Z (*Gen Z*; those born between 1997 and 2012) is experiencing the highest levels of climate anxiety. A global survey published in *The Lancet Planetary Health* in 2021 found that 75% of young adults ages 16–25 reported that the future felt frightening because of climate change (Hickman et al., 2021).

Other research suggests this anxiety is influencing key life decisions for Gen Z. Another study published in *The Lancet Planetary Health* focused on adolescents and young adults in the same age group in the United States. It noted that in the 3 years since the 2021 study the number of participants concerned about climate change had risen to 85%, with 64% believing it will affect their future choices, encompassing everything from where they live, to their career, to family planning (Lewandowski et al., 2024).

Figure 9.1 appeared in the 2023 Synthesis Report authored by the Intergovernmental Panel on Climate Change, which is part of

FIGURE 9.1. From Past Emissions to Future Lives: How Societal Choices Shape Intergenerational Climate Risk

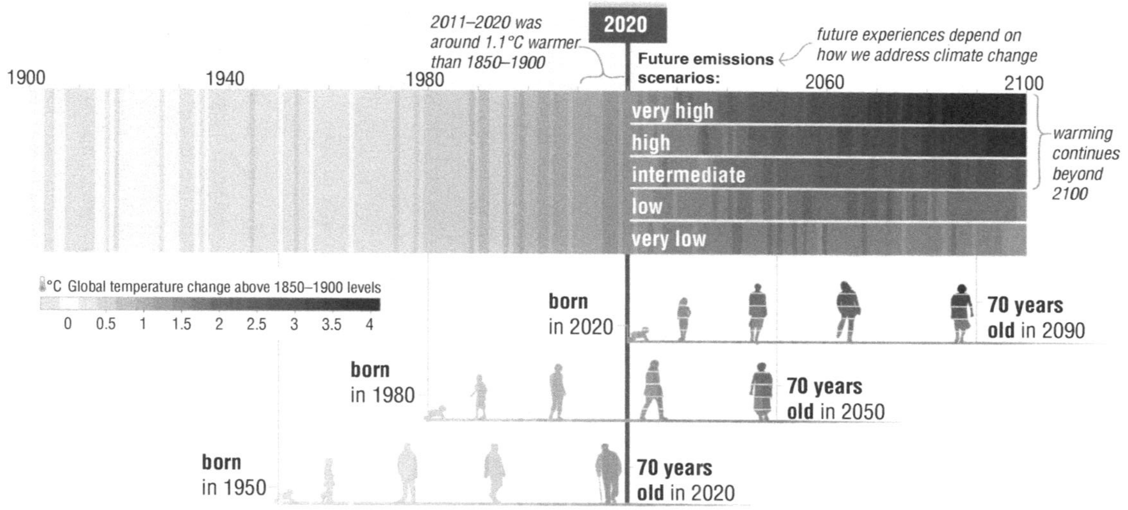

Note. The extent to which current and future generations will experience a hotter and different world depends on choices now and in the near term. Figure SPM.1 (c) in IPCC, 2023: Summary for Policymakers. In: Climate Change 2023: Synthesis Report. Contribution of Working Groups I, II and III to the Sixth Assessment Report of the Intergovernmental Panel on Climate Change [IPCC; Core Writing Team, H. Lee and J. Romero (eds.)]. IPCC, Geneva, Switzerland, pp. 1–34, https://doi.org/10.59327/IPCC/AR6-9789291691647.001. Use of IPCC figure(s) is at the User's sole risk. Under no circumstances shall the IPCC, WMO (World Meteorological Organization) or UNEP (United Nations Environment Programme) be liable for any loss, damage, liability or expense incurred or suffered that is claimed to have resulted from the use of any IPCC figure(s), without limitation, any fault, error, omission, interruption or delay with respect thereto. Nothing herein shall constitute or be considered to be a limitation upon or a waiver of the privileges and immunities of WMO or UNEP, which are specifically reserved.

the United Nations. The report illustrates how global warming is already affecting us and will be affecting future generations even more severely. It serves as a clear reminder of the responsibility we have to reduce greenhouse gas emissions in the years and decades ahead (Intergovernmental Panel on Climate Change, 2023, p. 7).

It is easy to understand why teens and young adults are feeling the most alarmed: They are facing the greatest consequences of the climate crisis, and many feel let down by their governments' lack of meaningful action. Their shock, confusion, anxiety, and anger are often fueled by a sense of betrayal because many of them have grown up in a world where the urgency of the situation has been apparent for years, and yet today's leaders are failing to confront the existential threat that their generation is inheriting. These emotions are only intensified when they hear well-meaning but out-of-touch comments from older generations: "It's inspiring to see how dedicated young people are to the climate crisis. You'll fix it." The sentiment that they are being "passed the torch" can feel dismissive to younger generations because, in contrast to a call for collective action from members of all generations, it shifts the burden onto the shoulders of those who feel they have been let down.

Isabelle Drury, a self-proclaimed "Zillennial" (born on the cusp of the Millennial and Gen Z cohort years), has described the relentless pressure she feels to stop the climate crisis:

> Like many of my fellow young people, I grew up with the inescapable knowledge that the climate crisis presented a very real risk to my future. . . . We have been awarded the unique experience of being called the last generation that can stop the climate crisis. . . . But it's actually incredibly demoralizing. It feels more like they're saying to me, "Yeah, I know we f* up the planet pretty bad (unknowingly and knowingly), but I can't do anything about it now. It's your problem. Good luck!" (Wray & Drury, 2023, para. 10).

Practice: Reflecting on Generational Perspectives

If you are a member of Gen Z, do you relate to the perspectives that Isabel Drury shared? What has your experience been with multigenerational support, or the lack of it? How do you feel about the weight of responsibility for change being placed on your generation?

If you are in a different age bracket, is there anyone in your life who is a Gen Z-er? If so, are you aware of how they are feeling? Have you had any conversations with them and listened to their experiences? Imagine feeling alone and unsupported as you grow up and face a rapidly warming future filled with unknowns. Wherever you are on the generational spectrum, take a moment to write about what it feels like to look toward an uncertain future, one shaped by increasing climate impacts and the reality that our efforts are falling short. As you reflect, keep in mind how these changes are already affecting today's children and youth. What emotions surface as you imagine what lies ahead, for yourself and especially for them?

INTERGENERATIONAL EFFORTS

> *Young people face the greatest consequences of climate change, but the responsibility for collective action falls to all generations.*

Young people face the greatest consequences of climate change, but the responsibility for collective action falls to all generations. Every generation brings its members own unique and essential insights, ideas, and skills. Whereas younger generations can introduce fresh approaches to solving problems, older generations often have more influence in political, economic, and institutional spaces. By working together, both groups can amplify their voices and push for meaningful change. Age diversity enriches the conversation and leads to more creative climate solutions.

Older generations were part of crucial social justice movements like the Civil Rights Movement, and they became masters of grassroots organizing. They used powerful tactics, such as peaceful protests, sit-ins, boycotts, and marches, to bring about societal change. The success of these movements in the 1960s often relied on building solidarity across different groups. Today's youth climate leaders can learn from these strategies about how to mobilize communities, build coalitions, and amplify their voices, because these time-tested tactics still apply in the context of climate advocacy.

Today, numerous influential youth-driven movements are making waves. Fridays for Future (https://fridaysforfuture.org/) has raised global awareness of climate change, spurred political action, and inspired millions to join protests demanding change (Fabel et al., 2022). The Sunrise Movement (https://www.sunrisemovement.org/; see also Sunrise Movement, n.d.) is pushing for the U.S. government to put a stop to the influence of fossil fuel elites, invest in minority and working-class communities, and create many more union jobs, ultimately putting ordinary people back in control. The Zero Hour Movement (https://thisiszerohour.org/) was founded to amplify the voices of young people of color and those on the front lines of the climate crisis. It also helped develop the Climate and Nature Bill introduced in the UK Parliament in 2020 and 2022 (Zero Hour, 2025). The group aims to tackle both climate change and biodiversity loss by creating a comprehensive framework to protect the environment and address global warming. These are just a few examples of how young generations are driving powerful initiatives for change.

Some of the most striking and successful examples of intergenerational efforts can be found in the youth-led legal battles for climate action. These cases have been driven by young people, but they have received substantial support from older lawyers, activists, academics, and organizations who understand the significance of intergenerational solidarity for tackling the climate crisis. Here are a few notable examples.

One of the most well-known youth-led cases is *Juliana v. United States* (filed 2015, dismissed 2020). Headed by the nonprofit organization Our Children's Trust, 21 young plaintiffs sued the U.S. government, arguing that its failure to act on climate change violated their constitutional rights to life, liberty, and property (Rush, 2025). The case was ultimately dismissed after the U.S. Supreme Court declined to hear the plaintiffs' appeal in March 2025, but it remains a landmark in both youth and environmental law. *Held v. State of Montana* (2024) saw a group of 16 young plaintiffs maintain that Montana's state policies were contributing to climate change and endangering their future well-being. This case became one of the first successful youth-led lawsuits filed against a U.S. state government over their climate policies. In August 2023, a district court ruled in favor of the youth plaintiffs, and in December 2024 the Montana Supreme Court upheld that ruling, ordering the state to implement stronger measures to reduce carbon emissions (Winters, 2025).

Similar efforts are taking place globally, including *Leghari v. Federation of Pakistan* (2015), *Urgenda Foundation v. The*

Netherlands (2019), *La Rose v. His Majesty the King* (2019), and *Youth v. Government of Mexico* (2020). These cases are more than just legal struggles: They are powerful demonstrations of the pivotal role young people are playing in the fight for climate justice and in holding governments accountable for the policies they create and the impacts they will have on future generations.

Practice: Exploring Intergenerational Efforts

Let's explore ways you can become a bridge-builder. Remember, your age does not matter: Reaching out to a different generation or social group begins with good communication. Use the deep listening techniques from Practice: Deep Listening in Chapter 6 to guide your approach. Start a discussion with someone from a different generation and explore how you can combine your unique gifts and talents. Keep in mind that you do not have to do something as formidable as launching a lawsuit: You can begin with a family, school, or community initiative.

Think about what you might need and what you can offer. Help in the form of academic or scientific information? Experienced lawyers? Mentorship or advocacy from climate activists or environmental organizations? Political or legislative support, or introductions to influencers? Financial funding, hands-on assistance, or perhaps artistic contributions, including help with website or social media design and text? Whatever you ask, or offer, take a moment to feel the gratitude and emotional connection that can blossom from these exchanges and the inspired action they can generate. You'll have a chance to write out a plan a bit later, in Practice: Writing for the Future.

STORYTELLING

Storytelling holds profound power. Take DearTomorrow, for example: Since its founding in 2015, this group has spent the past decade inspiring people to reflect on their climate legacy by writing personal letters to the future. As one of the first organizations to put into practice what Potential Energy Coalition's (2023) global report calls "generational urgency," DearTomorrow (n.d.-a) frames its theory of change this way:

> We're building the collective will to rewrite the next chapter of our climate story, one letter at a time. . . . Our project is a space to come to terms with complex emotions about climate change, reflect on our shared legacy, and build resilient communities that adopt positive lifestyles and behaviors. . . . Our love for the next generation compels us to rewrite the biggest story of our time. (paras. 1, 3, and 7)

But can writing letters really influence behavior? Research suggests it can be a surprisingly powerful way to inspire climate action—and one of the few strategies that bridges political divides. A 2021 study published in Climatic Change (Shrum, 2021), which has been supported by follow-up experiments (Syropoulos et al., 2023), found that although writing about climate risks did not alter people's perception of the risks, it did increase their willingness to take action (Vlasceanu et al., 2024). The greatest impact was seen among immediate family members: Grandparents who wrote to their future grandchildren donated 44% more than those who wrote to a family member without specifying a future recipient. It is important to note that political views made little difference: Even climate skeptics responded with increased generosity (Coren, 2025). Subsequent studies have reinforced these results. A 2024 "megastudy" published in the journal *Science Advances*, which tested 11 different interventions on more than 59,440 people across 63 countries, found that writing a letter to a future-generation member was the intervention most effective at boosting policy support (Vlasceanu et al., 2024).

DearTomorrow offers a variety of letter templates for different types of settings and groups. In the following exercise, let's try a basic one adapted from their website (DearTomorrow, n.d.-b).

Practice: Writing for the Future

Think of an important person in your life: your child, a friend, a family member, or your future self. Imagine that it is 2050, and they receive a message from you written today. Your message shares your thoughts about climate change and your promise to take action to ensure that they can live in a safe and healthy world.

Dear (insert name or relationship), ________________________.

In our beautiful world, I value and enjoy (describe a place or two that brings you inspiration or a sense of well-being) __________

____

____

____.

I'm writing to (express your concerns and feelings about the climate crisis) ____

____

____

____.

I'm hoping for a future that (describe your vision for a healthier, safer, cleaner world. What would it be like? Imagine and describe specific changes that would be in place) ______________________________

____

____

____.

I'm working to leave a stable climate legacy to you by (describe actions you're taking now or want to take soon to be part of making a positive change) ____

____

____

____.

With love,
(Sign your name with today's date) ________________________

You can also join more than 5,000 others and add your letter to the DearTomorrow digital collection by visiting https://www.dear tomorrow.org.

In the following exercise, you can take this a little further by pairing your letter with action steps.

Practice: Planning Intergenerational Action

Make a list of your gifts; include everything, from practical skills, financial assets, personal qualities, and abundant energy to talents and creativity and your various psychological resources.

____

____

____

____

____

____

____

Next, list your needs: practical help, financial assistance, specific talents or qualities, energy, creativity, and emotional resources.

____

____

____

____

____

____

____

Choose a person you know from a different generation, and imagine a conversation in which you explore aligning your gifts and needs with theirs. Be specific in your choices. Write about the potential benefits and outcomes of this partnership, considering how each person's strengths could benefit both people and the health of our Earth.

____

____

____

____

____

____

____

FAMILY PLANNING AND CLIMATE CHANGE

As climate change increasingly influences major life decisions of younger generations, one of the most deeply personal and complex dilemmas that young adults face is whether to start a family. A 2020 survey of 2,000 Americans commissioned by Seventh Generation and conducted by OnePoll found that 67% of Gen Z respondents felt that climate change had negatively affected their mental health, and 71% of millennials shared the same sentiment. Among 18-to-23-year-olds, four in five reported that they weren't planning—or didn't want—to have children as a result of climate change (McKibben, 2020). A global study published in *The Lancet Planetary Health* surveyed 10,000 individuals across 10 countries and found that 39% of teens and young adults expressed hesitation about having children because of climate change (Hickman et al., 2021).

> *One of the most deeply personal and complex dilemmas that young adults face is whether to start a family.*

In our warming world, the decision to have children is not just a personal one: It is shaped by the political choices made by our leaders, the health care that is available, financial security, job opportunities, and the projected safety of the environment in which the children will grow up. It involves ensuring that children won't be scarred by the impacts of climate change and the social instability it brings. It can be hard to find reassurance from our current reality and the climate trajectory being described by scientists.

Choosing to have no children, or fewer children than one had originally planned, is also recognized as a powerful form of climate action: A study titled "The Climate Mitigation Gap: Education and Government Recommendations Miss the Most Effective Individual Actions" (Wynes & Nicholas, 2017) gives voice to a growing and deeply personal concern among many parents and future parents: whether bringing a child into this world is compatible with caring for the planet. After analyzing carbon emissions, the authors, Seth Wynes, a Canadian researcher at the University of British Columbia, and Kimberly Nicholas, a researcher at Sweden's Lund University, found that having one fewer child is the single most impactful action an individual can take to reduce their carbon footprint—an insight that both validates and complicates the emotional landscape of climate-aware family planning.

Whatever you are feeling about the "kid question," you are not alone. The decision of whether or not to start a family in the face of climate change is not a one-size-fits-all question. Although fear and hesitation are understandable, there is a wide range of different perspectives, including the view that bringing new life into the world today expresses a deep belief in the potential of humanity's future. Young people can also feel pressure from friends, parents, or societal expectations that underscores the widespread lack of understanding of the challenges involved in making this decision:

- "So, when are you two planning to have kids?"
- "I've always wanted a grandbaby."
- "You've been married for a while now.... What's the hold-up?"

Some people find hope in bringing children into the world because they feel like it will make a positive difference by nurturing the next generation of caring, engaged humans who can help create a more sustainable future. The decision ultimately depends on our personal values, our belief in the potential for positive change, and our readiness to face the challenges that come with climate change. There are no right or wrong answers, and each person's decision will depend on their perspective, sense of responsibility, and their hopes for the next generation—all of which may change over time. Let's take a closer look at some of the concerns that emerge while reflecting on such a meaningful decision.

For people who are hesitant to have a child, or who have chosen to forgo having children, the decision often stems from an uncertainty about the future that has been influenced by topics from school or the news, including rising sea levels, extreme weather, food and water scarcity, and political instability. Others may worry about the strain on public health systems caused by air pollution, heat waves, and the spread of disease, questioning whether their country's health care infrastructure can keep up with the needs of an expanding population.

Raising children also requires considerable financial and emotional resources, and many are concerned that the increasing impacts of climate change and the resulting economic instability will make it more difficult to provide for a family. The ethics of the issue also weigh heavily on some people's minds: Is it right to bring children into a world so vulnerable to environmental collapse, to add to a growing population that will put even more pressure on Earth's resources and accelerating the environmental degradation already occurring? Some argue that having children in such uncertain times would be irresponsible because the future could bring even harsher living conditions that would be visited on future generations through no choice of their own. This has given rise to movements like BirthStrike (https://birthstrikemovement.org/), which was founded in 2018 by British musician Blythe Pepino and other young adults who declared they were refusing to procreate because they believed it would be morally wrong in the face of the climate crisis (McMullen & Dow, 2022). BirthStrike officially disbanded in August 2020. In 2019, 18-year-old Canadian McGill University student Emma Lim created #NoFutureNoChildren, a pledge to not have children until governments take serious action on climate change (McCall, 2019). Within weeks of its September 2019 launch, thousands of young people had taken the pledge, with reports citing figures ranging from 4,000 to more than 5,000 signatories by the end of that year.

Let's look at another side of what an excruciating decision this can be, starting with Britt Wray's story. Wray, PhD, is the founding director of the nonprofit group Unthinkable (https://www.unthinkable.earth/) and the founding director of CIRCLE, a research and action initiative at Stanford Psychiatry within the Stanford School of Medicine that focuses on Community-minded Interventions for Resilience, Climate Leadership, and Emotional Well-being. She has openly discussed her personal journey, which has included doubts about having children, and she has shared her exploration of ways to live more meaningfully in the face of the climate crisis.

Wray has explained that "eventually I did have a child, a boy," a choice she reached only "after more than three years of moral angst over the decision" (quoted in Schiffman, 2022, para. 24).

What helped guide her was the realization that "having a child was a commitment to joy despite living in a difficult time" and that to live a full life "you can't mute either the joy or the grief. The full human experience meant having a child" (Schiffman, 2022, para. 27). In another reflection, she wrote that what she sensed most deeply was a longing "to experience the deepest registers of love that humans are capable of feeling" (Wray, 2022, para. 10). She said she has wondered what challenges and surprises might arise from "a lifelong relationship with a little person we create" and whether the desire to provide a good life for that child might unleash "a new layer of our human potential" and a deeper sense of satisfaction with how she had lived (Wray, 2022, para. 10).

Wray joins those who believe that bringing children into the world is an act of hope and resilience. They have decided to embrace the challenge of raising a climate-conscious future generation by instilling values of conservation, responsibility, and sustainability. Although climate change presents serious difficulties, Wray and others believe that humans have the capacity to adapt and mitigate its damage with approaches that include technology, sustainability practices, and social change. And some advocates for reproductive justice argue that climate change should not strip individuals of their right to parenthood; instead, our efforts should concentrate on tackling the root causes of climate change, such as unsustainable consumption patterns, industrial pollution, and inequality, so that future generations can have the same opportunities to thrive.

Racial and cultural differences can also play a significant role in this decision. A 2023 study by Yale University and George Mason University revealed that Hispanic/Latino Americans were five times more likely (10% vs. 2%), and Black Americans were twice as likely (4% vs. 2%), to experience potentially serious levels of climate change psychological distress compared with their White counterparts (Uppalapati et al., 2023). Jade Sasser, PhD, an environmental scientist at the University of California, Riverside, has explored how climate change influences reproductive decisions, specifically examining racial disparities in attitudes toward starting a family in the context of environmental concerns. Her research sheds light on the intersection of climate change and reproductive justice by emphasizing how reproductive choices in marginalized communities are closely tied to larger social, economic, and environmental justice issues. In a 2024 interview with *Wired*, Sasser stated,

> Unfortunately, most environmentalists throughout the history of environmentalism have fallen on the side of saying don't have children. And I think that has been a very dangerous thing to say in particular, because those that they've been telling not to have children have tended to be low-income people and people of color. (St. Martin, 2024, para. 20)

Sasser has also studied how racial and cultural values shape attitudes toward parenthood. In some communities, there may be strong cultural or familial expectations to have children despite environmental concerns; in others, climate issues may weigh more heavily in the decision to delay or forgo parenthood. One of the most striking findings in her research is that women of color reported being most likely to have fewer children than they desired because of climate change. "No other group in that survey responded that way," Sasser noted (St. Martin, 2024, para. 6).

Whether you are personally grappling with this topic, supporting someone who is, or are broadly exploring key climate change themes, clarifying your views on this highly charged issue is a crucial aspect of climate literacy. Although we want to process all the emotional aspects of sensitive climate topics, we also want to analyze situations from multiple angles, think deeply about them, and make decisions that are well informed. The goal of the following exercise is not to arrive at an answer but rather to foster an open and ongoing reflection.

Practice: Reflection on Children and Our Future

1. Can you identify any biases that might influence your views? (See Chapter 5.)

2. What are the potential consequences (both intended and unintended) of the actions that you or others might take in this regard?

3. How does this issue connect to broader societal, historical, or cultural contexts?

4. What are the ethical implications of this decision, and how could different ethical frameworks lead to different conclusions?

5. How can empathy for and understanding of opposing viewpoints help you develop a more nuanced perspective on this topic?

6. In what ways does this concern affect different groups of people, and are some voices being marginalized or ignored?

7. How can we balance the short- and long-term consequences of our decisions about child-bearing in the era of climate change?

____

____

____

____

____

____

LOOKING FAR ENOUGH AHEAD

It is crucial to shape our environmental care strategies with future generations in mind—but are we truly looking far enough ahead? As American writer and environmentalist Terry Tempest Williams (2002) poignantly reminds us, "The Eyes of the Future are looking back at us and they are praying for us to see beyond our own time" (p. 99). The following exercise takes inspiration from the Haudenosaunee (Iroquois) Confederacy, a powerful alliance of six Native American nations: Mohawk, Oneida, Onondaga, Cayuga, Seneca, and Tuscarora (National Museum of the American Indian, Education Office, 2009). Their Great Law of Peace (*Gayanashagowa*) is an ancient system of governance, ethics, and social organization that is recognized as one of the world's oldest participatory democracies. It is known for the principle of *Seven Generations Thinking*. In simple terms, this means that when making decisions we need to consider how they will affect the well-being of the next seven generations (roughly 150 years). This approach ensures that our choices will respect the past, support the present, and preserve the future.

> *It is crucial to shape our environmental care strategies with future generations in mind—but are we truly looking far enough ahead?*

The following exercise encourages you to focus on what truly matters and how you can live a meaningful life that is based on the values you deem important. It can help you prioritize your relationships, passions, and personal growth and ultimately clarify your contribution to the ongoing well-being of life on the planet.

Practice: Your Great-Great Grandchild's Story

Let your imagination transport you 200 years into the future, where you are invisibly peering into a family's living room. You see your great-great-great-great-great-grandchild—or a child who is connected to your family, or a close friend's heritage, in some way—sitting with other family members around the kitchen table. The child says to the others, "Tell me the story about [your name] again." What would you want that child to learn about you? This is a powerful practice that helps you reflect on what truly matters in your life and the legacy you would like to leave behind. Here are some considerations:

- How would you want others to describe who you were?
- What qualities, achievements, or relationships do you want to be remembered for?
- What values, passions, or actions would you like to define your life?
- What impact did you have on the people around you?
- Which of your personal traits—such as kindness, creativity, or courage—would you want them to associate with you?
- Did you make a meaningful contribution through community service, leadership, or love?
- How did you live your life: fearlessly, humbly, with passion?

- Write the story that your great-great-grandchild (or another young person) is being told around the table.

Once you have written the story, think about your current life with respect to how you would like to be remembered. Is there anything you need to change in your life to make that ideal legacy a reality?

In Chapter 10, the final chapter of this book, you will learn more ways to get involved in tackling climate change with guidance from one of the world's top scientists. You will find more practice exercises that focus on resilience, and we will integrate and synthesize the perspectives you have read about throughout this book. You will discover how your engagement can lead to the discovery of true hope in the face of the climate crisis—not as naive optimism but as a sustainable commitment that keeps your passion for action alive.

CHAPTER 10

WE'RE THE ONES WE'VE BEEN WAITING FOR

As you have progressed through these chapters, you have immersed yourself in the profound challenges of climate change, exploring its complex realities while learning to navigate the emotions it stirs. You have been cultivating resilience so you can face the unknown with courage and empathy. Ultimately, doing this is about showing up in moments of uncertainty and disruption with beauty, heart, and soul to a world that is in need of both care and clarity. As you face this future, you are not doing it alone—you are participating in the essential work of building community, strengthening bonds across generations, and widening the circle of people who can hold and shape this moment with you. Let's start this chapter by engaging in a practice exercise that weaves together these different elements and integrates them into a grounded and cohesive experience.

The following exercise is a meditative and reflective activity that is most easily experienced when you are guided by the script. It invites you to feel, from the inside out, that your very being is woven from the same air, water, and ancient minerals that have always moved through this Earth and always will, carrying you into the timeless flow of ancestral connections that live within and around you. You can practice with a friend and take turns reading the script aloud, or you can record it and listen to it when you are ready to start. This is also a beautiful practice to share with a group. Remember to read slowly and leave short pauses between paragraphs.

Practice: We Are Nature

Find a quiet space and set aside 20 minutes when you won't be disturbed, making sure to turn off your phone. It is best to practice while sitting in a chair or on the floor. Adjust your posture as needed so that you feel supported and your breath can flow freely.

Close your eyes or give them a soft downward focus. Take a few full breaths: Breathe in, and, as you exhale slowly, soften your forehead . . . your eyes . . . your jaw . . . and your shoulders. Let go of weight or tension you may be carrying: Imagine and feel your shoulders relaxing down away from your ears. Soften your hands . . . your belly . . . your hips . . . and your feet.

Feel yourself—your whole body and whole being—being present here and now.

Now bring your awareness inside to the center of your pelvis, visualizing a small ball just below your navel that's evenly centered between the sides, front, and back of your body. With your mind's eye, draw a straight line from that ball straight down to the center of the Earth, and feel them connected by the gentle pull of gravity. Imagine the energy of your breath starting at the center of the Earth, rising up into the front of your body as you inhale, rinsing up over your scalp, and flowing down the back of body as you exhale, returning into the Earth. Continue this for several cycles of breath for about 2 or 3 minutes.

Now gently bring your focus to the memory of what you have eaten for the past day or two and how those meals have become part of your body and are present in you right now. Consider the origins of one of the fruits or vegetables you have eaten, how during their lifespan they absorbed sunlight, water, and minerals from the soil.

The minerals that make up the rocks in the area where they thrived are now part of your bones. The water that was once in a river or rain now flows through your veins. Imagine how the fruit or vegetable was carried by many different hands during the planting, picking, packing, and distribution activities that eventually made them available to you. Feel the connections. . . . Feel all that made it possible. Feel gratitude.

Bring your attention to the air above you and around you. When we draw air into our lungs, we call it breath but notice how there is no actual division from the sky. Breathe the sky and welcome it into your body. Breathe the wind that mingles with the plants, that sustains all of life. Feel the connections. Feel gratitude.

Tune into the rhythm of your breath as it flows in and out of your body. Notice how your ribs and belly expand with each inhale and soften with each exhale—filling up, emptying out, over and over. This is an ancient rhythm that is like the ebb and flow of tides, the waxing and waning of the moon, and the cycles of the seasons. Feel life, and feel nature as it breathes you. You are woven into the pulsing rhythm of existence. Embrace the connections. Feel gratitude.

Feel this present moment as part of a continuous flow. Ancestral influences before you have guided you here, while around you relationships form circles with layers of meaning, emotion, and support. Ahead, the journey continues rippling out to future generations. Recognize your role in this unfolding expression of life, both unique and shared. Feel the connections. Feel gratitude.

Before concluding this practice exercise, take a moment to notice any changes in your body, mind, emotions, and sense of connection to the natural world. Linger a little longer, allowing any shifts to settle in as a new template for your expanded sense of self. It is not just your mind, but also your body, nervous system, and emotions, that will remember.

As you are ready, begin to sense the room around you, your presence in your body here and now . . . and gently open your eyes.

Write about your experience. Did any aspects feel familiar? Did anything surprise you? What qualities are you most aware of now? Is there anything from this practice exercise that you would like to nurture regularly and bring into your everyday life?

____

____

____

____

____

When we move through life with an active awareness of our deep interconnection with nature, we begin to become more fully attuned to how our choices ripple across human and nonhuman communities, now and into the future. We are reminded that both outside our windows and within our beings, everything is cycling and flowing, offering its support and asking for ours. This awareness invites us to reckon with the true costs of how we live today and to more intentionally imagine and shape a safer, more life-honoring future.

When we navigate life with an active awareness of our inseparable connection to nature, we become more attuned to how our actions affect every member of the countless communities of humans, nonhumans, and future generations. This awareness helps us recognize the true costs of our current ways of living and more thoughtfully explore how to cocreate a safer future.

THE POWER OF "WE"

Although you will find your unique ways of engaging in change, it is also important to consider guidance from experts in the field of climate science. One of the world's best is Katharine Hayhoe, PhD, who holds many roles and distinctions. She is chief scientist at The Nature

Conservancy and a Paul Whitfield Horn Distinguished Professor, and holds the Political Science Endowed Chair in Public Policy and Public Law in the Department of Political Science at Texas Tech University. She is also a principal investigator for the U.S. Department of the Interior's South Central Climate Adaptation Science Center (https://www.usgs.gov/programs/climate-adaptation-science-centers/south-central-casc) and works with the National Science Foundation's Global Infrastructure and Climate Network (https://www.icnetglobal.org/about/). Hayhoe is frequently asked what we can do about climate change, and for answers she draws from research conducted in the physical and social sciences. She says that the following interventions will be the most effective for propelling us toward the positive social tipping points essential for a rapid worldwide carbon-neutral transformation (Otto et al., 2020):

1. Remove fossil fuel subsidies and incentivize decentralized energy generation
2. Build carbon-neutral cities
3. Divest from assets linked to fossil fuels
4. Reveal the moral implications of burning fossil fuels
5. Strengthen climate education and engagement
6. Disclose greenhouse gas emissions information

Hayhoe (Otto et al., 2020) notes the strong role for individuals in numbers 3, 4, and 5, and she reminds us that the only way that numbers 1, 2, and 6 will happen is if individuals also push, advocate for, and demand these changes at every level of governance and society. She echoes the sentiment of Bill McKibben (2017), Schumann Distinguished Scholar in Environmental Studies at Middlebury College and founder of Third Act and 350.org, who has stated that, "The most important thing an individual can do right now is not be such an individual" (para. 14). Collective action aligns with the African proverb "If you want to go fast, go alone. If you want to go far, go together"—a powerful reminder that the power of community and collective action is a value with deep and global roots, woven through cultures around the world. Hayhoe challenges either–or thinking with a "yes–and" mindset: "Do we need individual or systemic change? YES! Because what is a system made of if not people, and how does it change if not when people make it?" (Hayhoe, 2024).

> *Collective action aligns with the African proverb "If you want to go fast, go alone. If you want to go far, go together"*

Hayhoe (2021) highlights the importance of starting conversations about why climate change matters and what individuals can do to make a difference. She encourages everyone to join a climate action group in order to amplify our voices and make our actions contagious by sharing them with others. She also suggests reviewing where we keep our money—including our bank, credit cards, retirement accounts, and investments—and using our voice to push for change. Begin to spark ideas for change at work, school, or any group you belong to, whether it is a congregation, a book club, or a sports team. Hold politicians accountable at every level, from city to regional and state officials, because they can sometimes make swift decisions that have a significant impact. She notes that they often underestimate how concerned their constituents are about climate change because they hear from them so rarely.

THE CASE FOR HOPE

Given her deep scientific knowledge and understanding of social behavior, is Hayhoe hopeful about the state of the world? The answer lies in the title of her book, *Saving Us: A Climate Scientist's Case for Hope and Healing in a Divided World* (Hayhoe, 2021). She acknowledges that it is easy to feel defeated with climate crisis stories making the headlines every day—but she reminds us that there are many examples of hopeful stories from companies, governments, people, and organizations taking action that get lost amid the doom and gloom.

In Chapter 5, we explored a number of cognitive biases, and there is another one that is especially relevant to our discussion about hope: *negativity bias*. Like other biases (see Chapter 5), it evolved from an early survival mechanism that helped us quickly detect and identify threats and remember negative experiences in order to enhance our chances of survival. With the negativity bias, our brains are wired to give three to five times more attention to negative stimuli (e.g., threats and fears), and the feelings they elicit tend to remain with us longer than feelings and sensations that are neutral or uplifting (Hanson, 2013; Rozin & Royzman, 2001).

Climate change is deeply concerning, and the negativity bias serves to amplify our anxiety. We should recognize that the negativity bias does not diminish the urgency or importance of addressing climate change, but it does emphasize the need for us to be more intentional in focusing on positive progress and solutions so that we maintain a realistic outlook. For balance, we need to keep an eye on the people and organizations that are making real impacts in the most critical areas. Here are a few examples from around the world:

- California has implemented a cap-and-trade system that sets a limit on total emissions and allows companies to buy and sell allowances for emissions, incentivizing businesses to reduce their carbon output (Center for Climate and Energy Solutions, n.d.).
- Sweden introduced a carbon tax in 1991, and it has been one of the most effective policies for reducing emissions while also maintaining economic growth (Government Offices of Sweden, 2024). Sweden now has one of the lowest carbon emissions per capita in Europe (Andersson, 2019).

- The Rodale Institute in the United States has been pioneering regenerative agricultural practices, such as no-till farming, cover-cropping, and rotational grazing, to increase soil health and sequester carbon (Moyer et al., 2020).
- Bogotá has invested in the TransMilenio system, a high-capacity rapid transit system that significantly reduces traffic congestion and greenhouse gas emissions in the city and encourages sustainable urban mobility (Institute for Transportation & Development Policy, 2024).
- Companies like Gotham Greens in the United States practice hydroponic farming, which involves growing crops in water-based, soil-free systems. This method uses up to 90% less water than traditional agriculture, making it ideal for water-scarce areas (Marks, 2024).

Hope in the context of climate change is not a naive optimism that passively waits and thinks, "Someone will figure this out, they always do." It's the kind of hope described by David Orr, PhD, Paul Sears Distinguished Professor of Environmental Studies and Politics Emeritus at Oberlin College, who defines it this way: "Hope is a verb with its shirtsleeves rolled up" (quoted in Weinstein, 2021, para. 5). This is different from optimism and an important distinction given the many uncertainties and unknowns surrounding the future. Orr explains that authentic hope demands that we set aside blind optimism and confront the future with honesty, self-awareness, and a clear-eyed understanding, qualities that are often difficult to sustain. In essence, active hope embodies the ability to hear the calls to action around us while standing in the unknown.

> *"Hope is a verb with its shirtsleeves rolled up"* (quoted in Weinstein, 2021, para. 5).

SHIFTING NORMS THROUGH TRUSTED RELATIONSHIPS

You may wonder, as I often have, whether there are enough people involved in this kind of action-oriented hope to create the positive impacts we yearn for. How can we get everyone on board? The great news, according to Damon Centola, is that we don't need everyone: Experimental research shows it takes only about 25% of people in a social system to shift norms and ignite widespread change, though researchers note the exact threshold can vary depending on the circumstances (Centola et al., 2018). Centola, PhD, is the Elihu Katz Professor of Communication, Sociology, and Engineering at the University of Pennsylvania, where he directs the Network Dynamics Group. In a study published in *Science*, he and his colleagues demonstrated that when roughly 25% of individuals in a network adopted a new behavior or norm, this created a tipping point that led the entire population to eventually shift toward it (Centola et al., 2018).

Centola et al.'s (2018) research shows that traditional strategies, like announcements and awareness campaigns, usually are not enough to create true, lasting change. Instead of spreading like information, change moves through trust and social connections, a process he calls *complex contagion* (Centola, 2018). At the heart of this theory is a simple but powerful idea: Real change happens when people are exposed multiple times to a new behavior, especially by those they trust.

> *Real change happens when people are exposed multiple times to a new behavior, especially by those they trust.*

In other words, it is not just the message itself that drives change: It is who is connected, to whom, and how those connections are structured, that matter most. People rarely shift their behavior simply because a leader tells them to or because they hear about something once. Instead, change takes root when they repeatedly see trusted peers adopting new behaviors. As these exposures build over time, the new behavior starts to feel familiar, credible, and real, and eventually gains the momentum needed to spread across the network.

EXISTENTIAL INQUIRY

Climate change has been described as an *existential crisis*, and so we can also respond to it through existential inquiry. Cultural anthropologist Angeles Arrien shared that the tribal peoples of South Africa teach that from the moment we are born, Death walks on our right and Destiny on our left. Each day, Death asks, "Are you using the great gift of life well?" while Destiny asks, "Are you doing what you've come here to do?" (Arrien, 2012, para. 2). Let's approach climate action from this deeper perspective, which understands that addressing climate change is not just about fixing the problem: It is about embracing the transformative vision that has guided your journey throughout this book and that is unfolding in your life.

When you hear the whispers of Death and Destiny, reflect and consider how well you are using your gifts, talents, and time. Set aside any external pressures or rigid expectations, and take time to contemplate and consciously choose to stay true to your deepest sense of self while working toward meaningful goals. Trust that you can align your actions more closely with what you truly care about and believe in. Find the place within that stirs you to keep going through the heat and the rains, through the weight and the restoration. "Emergency" and "emergence" share the Latin root *emergere*, meaning "to arise, bring to light, or bring forth" (Harper, n.d.). In the midst of disruption, we are called to rise: to make visible what has been hidden, and to meet the moment as it unfolds.

A NEW CLIMATE ACTION PLAN

Practice: Revising Your Climate Action Plan focuses on refining the commitments you've already identified. This might involve refining what you are already doing, perhaps by adding more support or involving others. Or you may feel ready to step outside your comfort zone and explore something new. It could also mean scaling back to create more emotional balance in your life to ensure that you can stay engaged and energized over the long run. It is about creating or updating a climate action plan on any level, whether that involves changes at home, school, or work, in your community, or on the world stage.

In Chapter 5, we began exploring how to engage in climate action while considering the bystander bias. In the chapters that followed, you have delved into many other aspects of climate emotion and engagement, and Practice: Remembered Strength gives you an opportunity to revisit those initial thoughts and decide whether you want to strengthen or adjust them. Your plan doesn't need to be focused on climate change, as such: Your efforts may be more broadly focused on helping bring people together to create a healthier and safer world. Your plan might be about making the air and water healthier in your community; reducing plastic pollution to protect marine life; working on regenerative farming practices; promoting local, food, or supporting agroecology; or bringing social/emotional learning curricula into local schools.

Reflect on the themes you have explored so far throughout this book. In particular, consider the following:

- **Pleasure activism:** Remember this concept from Chapter 1? Ensure that your climate-related activities themselves bring joy so that the process is part of the reward.

- **Existential values:** Are your actions aligned with your clearest sense of self and informed by the whispers of Death and Destiny? Are you both cultivating and offering all of your unique talents and gifts?
- **Impact on others:** Who benefits from your actions, and who might be left out? Consider inclusivity in your approach. Will it help build connections and community?
- **Social justice:** Does your plan include a social justice component? How does it address fairness and equality?
- **Long-term impact:** How might your efforts affect the next seven generations? Think about the legacy you are creating.
- **Emotional support:** How can you take care of your emotional well-being amidst the challenges of this work? Think about your strategies for maintaining resilience and balance.
- **Transforming biases:** Pay attention to the stories you tell yourself, your beliefs about who you are, or your recurring thoughts about a specific situation or life in general. As you become more aware of your patterns and biases, are there any revisions you would like to make to ensure that they express a more accurate or empowering truth?
- **Creativity and adaptability:** How can you keep your approach fresh and creative and stay flexible in how you are moving forward?
- **Commitment:** When and how will you begin? What specific steps will you take to start or continue your journey of meaningful change? Are there obstacles you anticipate? If so, how can you address them?
- **Support:** Consider the forms of support you may need—people, resources, or personal strategies—in order to stay on track and remain motivated.

Practice: Revising Your Climate Action Plan

After reflecting on the themes in the bulleted list provided earlier, write about how you want to participate in meaningful change, whether by taking new steps or refining your current commitments. Be specific about how you can incorporate into your actions the essential elements in that list.

INNER STRENGTH

The following exercise, the final one in this book, will help you tap into an inner strength you may not always know is there. It will empower you to face the challenges ahead by guiding you to reflect on past difficulties and uncover the skills or wisdom you have gained from them. Growing from tough experiences is not about minimizing the pain you have endured; instead, you can recognize how those experiences have shaped you and how you can use that growth to create a better future. Perhaps you have learned what truly matters to you, have begun to set clearer boundaries, or developed a deeper sense of empathy and compassion for the struggles of others. This self-awareness allows you to better understand your strengths, weaknesses, and triggers and provides valuable insight for making decisions and nurturing relationships as you move forward with your action plan.

Practice: Remembered Strength

Reflect on a time in your life when you faced a significant challenge. There may be several that come to mind, so choose one that you emerged from stronger—or perhaps you are still in awe of how you made it through. If nothing specific comes to mind, imagine how you would like to approach future adversity: What qualities would you like to draw on from within yourself and with the support of others?

Find a comfortable seat, and close your eyes or gaze downward with a soft focus. Take a deep breath in through your nose, feeling your belly expand as you inhale, and then exhale slowly through your mouth while inviting your body and mind to relax. Repeat this relaxing breath two or three more times, continuing to let go of tension and stress as you settle into the present moment.

As you continue to breathe deeply and calmly, allow your mind to drift back through time, and revisit the memory of the challenge that came to mind. If a different moment surfaces spontaneously, feel free to explore it, or gently return to the challenge you initially intended to reflect on. Let the difficult experience come into focus.

As you revisit this challenge, you may feel the pull to linger on what was painful or overwhelming—that's understandable. For this practice exercise, though, let the spotlight rest on what carried you through. Focus on the strengths that rose up in you, even quietly: your determination, your patience, your courage, the way you let your emotions move, or how you simply took things one step at a time and trusted your instincts. Maybe you reached out for support. Whatever helped you navigate that moment is what we are honoring here. Take a moment to vividly recall how you found your footing during that time. Step into the qualities of strength that helped you move though the situation, and allow that strength to fill your body, mind, and emotions. What did you see, hear, sense, and feel? Let yourself feel that resilience growing stronger, spreading like a warm wave of confidence throughout your entire being. Know that this strength is always within you: It is a part of you that is ready to be called on whenever it is needed.

Imagine carrying that strength within you from the past into the present moment. Visualize bringing it to a new challenge you are facing, or to one that might occur in the future. Take a deep breath and allow the strength you just recalled to continue flowing through you. See yourself moving through this new challenge one step at a time as you are guided by wisdom and strength you have already built. Trust that you have the ability to tap into this inner strength whenever you need it—and stay humble, knowing that more learning lies ahead.

When you are ready, gently bring your awareness back to the present moment. Begin by noticing the sensations in your body and the sounds around you. Slowly wiggle your fingers and toes and, when you are ready, open your eyes and feel the sense of strength and confidence that is with you now. Take a few moments to write about your experience.

____

____

____

____

____

FINAL THOUGHTS

Our deep caring and concern about the most pressing challenges of our time—climate change, social justice, public health—come from a place within us of profound love and empathy. This love is not about fleeting sentiments; it is a steadfast commitment to what matters most: our children, animals, cherished places and communities, and the very essence of life itself. It is a love that drives us to protect, preserve, and nurture—not just for today but for future generations as well. This love fuels our urgency to act, to create meaningful change, and to ensure that we build a world that is in alignment with the laws of nature and that is just, healthy, and sustainable for all.

When progress is slow, setbacks seem endless, or when things don't go as planned, it can feel disheartening. The weight of what is at stake—the lives, the ecosystems, and the future of our planet—can sometimes feel like too much to bear. It is easy to question whether our efforts are enough, or if we will ever see the changes we so desperately need. But right now you are more equipped than ever to be resilient, empathic, and clear minded and to join the growing movement of people who are waking up, ready to do their part in what is required for planetary healing. We can all cultivate our capacities for repair and restoration and help everyone feel heard, valued, and safe. Our care is an unshakable force.

To all those who bring their heart, dedication, and unwavering resolve to work for a healthier and more just world: Thank you! Let's remind each other that the pursuit of a better world is always worthwhile. Your vision of a future where the Earth, its people, and its ecosystems thrive is part of the collective inspiration that keeps us connecting and working toward transformation, even when the path ahead is tough. As a message attributed to the Hopi Elders (2000), and reflected in Alice Walker's (2006) book and speeches at the World March of Women, remind us: "We're the ones we've been waiting for" (p. 3).

APPENDIX

CLIMATE RESILIENCE RESOURCE GUIDE

This guide highlights a curated selection of organizations and media designed to help you strengthen resilience in the face of climate challenges. New resources are emerging constantly across every category. The resources included here span both *inner capacities*—emotional and mental well-being—and *outer capacities,* such as action, advocacy, and community engagement.

This guide is not meant to be comprehensive but rather provide a meaningful starting point. If you don't find what you need, try exploring local offerings—or even consider creating the resource you wish existed.

SUPPORT FOR CLIMATE-RELATED EMOTIONAL DISTRESS

Individual Support

Climate Psychology Alliance (North America): https://climatepsychology.us

Directory of climate-aware therapists trained to support individuals experiencing climate-related distress. Their main site also includes a directory of "climate cafés," monthly virtual gatherings to reflect and connect around the emotional dimensions of the climate crisis as well as information about trainings and lectures on climate distress for mental health professionals.

Climate Psychology Alliance (International): https://climatepsychologyalliance.org

International provider listings (including the United Kingdom, Spain, Australia, and Italy). New practitioners are added regularly. The site also offers groups and additional support resources.

Group Support

The All We Can Save Project: https://allwecansave.earth

Encourages small-group dialogue to explore the *All We Can Save* anthology (https://www.allwecansave.earth/anthology) and help cultivate climate leadership.

Johnson, A. E., & Wilkinson, K. K. (Eds.). (2020). *All we can save: Truth, courage, and solutions for the climate crisis*. One World.

Good Grief Network: https://www.goodgriefnetwork.org

Provides information related to peer support groups for processing eco-anxiety and building resilience.

One Resilient Earth: https://www.oneresilientearth.org

Offers information about creative workshops and group spaces that cultivate resilience and regenerative action.

EDUCATION AND RESOURCES ON CLIMATE MENTAL HEALTH

Climate Mental Health Informational Sites

American Psychological Association, Climate Topics: https://www.apa.org/topics/climate-change

Offers links to podcasts, reports, and other climate-related mental health resources.

Climate & Mind: https://www.climateandmind.org

An interdisciplinary hub that explores the psychological dimensions of climate change.

Climate Mental Health Network: https://www.climatementalhealth.net

Provides resources for professionals and communities on climate-related mental health.

Climate Psychiatry Alliance: https://www.climatepsychiatry.org

Psychiatrist-led organization that focuses on developing public and professional climate mental health resources.

Ecopsychepedia: https://www.ecopsychepedia.org

Provides a free online glossary and knowledge hub for psychology–ecology–climate terms.

Unthinkable: https://www.unthinkable.earth

A resource hub that supports public conversation on climate and mental health.

Yale Program on Climate Change Communication: https://www.climatecommunication.yale.edu

Provides information about research on climate beliefs, attitudes, policy preferences, and communication strategies.

Climate Psychology Training

If this book sparks interest in pursuing climate psychology, or weaving psychological insight into your current work, these programs can help you deepen your path.

Climate Psychology Certificate, California Institute of Integral Studies:
https://www.ciis.edu/public-programs/climate-psychology-certificate

This is a university-based professional certification; sessions are offered live online with international faculty.

Workshops through the Climate Psychology Alliance: https://www.climatepsychologyalliance.org

Trainings and lectures on climate distress for mental health professionals are available.

SUPPORT FOR KIDS, FAMILIES, AND EDUCATORS

Support for Children

American Psychological Association Children's Workbooks

- *All the Feelings Under the Sun: How to Deal With Climate Change* (ages 9–13): https://www.apa.org/pubs/magination/all-feelings-under-sun
- *What to Do When Climate Change Scares You: A Kid's Guide to Dealing With Climate Change Stress* (ages 7–11): https://www.apa.org/pubs/magination/what-when-climate-change-scares-you

Children's Environmental Health Network: https://www.cehn.org

Offers research, policy, and education to protect children from environmental harms.

Support for Parents and Families

The Healing Library: https://www.thehealinglibraryus.com

Supplies free lending kits to help families process difficult topics, including environmental grief.

Moms Clean Air Force: https://www.momscleanairforce.org

A parent-led group that advocates for clean air and climate justice.

Our Kids' Climate: https://www.ourkidsclimate.org

A global coalition of parents advocating for climate-safe futures.

Support for Educators

Climate Mental Health Network education resources: https://www.climatementalhealth.net/education

Provides tools for K–12 educators who are supporting students experiencing climate anxiety.

Climate Psychology Alliance guides for educators: https://www.climatepsychology.us/resource/an-educators-guide-to-climate-emotions/

Provides free resources for teaching climate science and emotional resilience are available.

Talk Climate: https://www.talkclimate.org

Offers resources to encourage constructive climate conversations in classrooms and communities.

National Environmental Education Foundation: https://www.neefusa.org

A leading lifelong environmental education organization.

CORPORATE AND ORGANIZATIONAL SUPPORT

Project InsideOut (Renée Lertzman): https://reneelertzman.com/project-inside-out

Project InsideOut offers psychology-based tools for to facilitate climate engagement within organizations.

The Adaptive Mind Project: http://susannemoser.com/documents/AdaptiveMindOverview5-27-19.pdf

The aim of The Adaptive Mind Project, which is led by Susanne Moser, is to help climate and justice practitioners build emotional resilience and leadership capacity.

SUPPORT FOR ARTISTS AND CULTURAL PARTNERS

Artists & Climate Change: https://www.artistsandclimatechange.com
Provides a directory of organizations working at the intersection of arts and climate.

The Climate Museum: https://www.climatemuseum.org
The first U.S. museum dedicated to climate through art, culture, and community.

Environment & Culture Partners: https://www.ecprs.org/
Connects cultural institutions with environmental organizations.

Hollywood Climate Summit: https://www.hollywoodclimatesummit.com
Annual convention of entertainment professionals for the purpose of advancing climate storytelling.

GENERATIONAL AND CROSS-GENERATIONAL RESOURCES

Youth-Led Climate Action

Extinction Rebellion Youth: https://rebellion.global/
A youth-led branch of Extinction Rebellion that uses nonviolent direct action to demand urgent, systemic responses to the climate and ecological crises.

Fridays for Future: https://www.fridaysforfuture.org
An international youth movement that mobilizes young people worldwide to demand science-based climate action and justice.

Friends of the Earth International: https://www.foei.org
A global federation of grassroots environmental organizations working collectively for environmental justice, human rights, and sustainable societies.

Sunrise Movement: https://www.sunrisemovement.org

A U.S.-based youth-led organization focused on building political power to advance climate action and economic justice.

Cross-Generational Action Groups

350: https://www.350.org

A global grassroots climate movement.

The Climate Reality Project: https://www.climaterealityproject.org

Provides resources for climate leader training.

Earthjustice: https://www.earthjustice.org

Focuses on legal advocacy for climate and environmental protection.

Citizens' Climate Lobby: https://www.citizensclimatelobby.org

Provides information related to grassroots political advocacy.

The Nature Conservancy: https://www.nature.org

Offers information related to nature-based climate solutions.

Organizations for Older Adults

AARP Climate Initiatives: https://www.supportaarpclimateaction.org/

Provides climate resilience and clean energy for older adults.

Third Act: https://www.thirdact.org

Mobilizes people over age 60 for climate justice and democracy.

FAITH COMMUNITIES WITH AN ENVIRONMENTAL FOCUS

GreenFaith: https://greenfaith.org/

Focuses on interfaith climate justice organization.

Interfaith Power & Light: https://ipldmv.org/

Located in the DC, Maryland, and Northern Virginia region; provides information on renewable energy and climate action for congregations.

One Earth Sangha: https://www.oneearthsangha.org

Buddhist-inspired ecological practice and activism.

Parliament of the World's Religions: https://www.parliamentofreligions.org

Global spiritual communities addressing climate challenges.

United Religions Initiative: https://www.uri.org

Grassroots peace, justice, and environmental projects.

Faith for Earth Coalition: https://www.unep.org/about-un-environment/faith-earth-initiative

An initiative of the UN Environment Programme that partners with faith groups on environmental protection.

CLIMATE SCIENCE RESOURCES

Intergovernmental Panel on Climate Change: https://www.ipcc.ch

Conducts global scientific assessments of climate change.

National Aeronautics and Space Administration: https://science.nasa.gov/climate-change/

A source of climate data, visualizations, and research.

National Oceanic and Atmospheric Administration: https://www.noaa.gov

A source of climate data and forecasts.

Project Drawdown: https://www.drawdown.org

Offers information about science-based climate solutions.

MEDIA AND NEWSLETTERS

Grist: https://www.grist.org
Journalism that addresses independent climate solutions.

Inside Climate News: https://www.insideclimatenews.org
Pulitzer Prize–winning clean energy and climate reporting.

Drilled (Podcast): https://drilled.media/podcasts/drilled
Investigative reporting on barriers to climate action.

What If We Get It Right? (Podcast): https://www.getitright.earth/podcast
Hope-forward, solutions-oriented climate storytelling.

Britt Wray's *Unthinkable Times* blog: https://unthinkable.substack.com/
Essays and resources on climate emotions.

Living on a Changing Planet (Podcast): https://www.livingonachangingplanet.com
Focuses on intersections of climate and mental health.

FILMS AND LITERATURE

Films

2040 (2019): https://theregenerators.org/2040/
Provides an optimistic vision of solutions-based climate futures.

THE HERDS (2025): https://www.theherds.org
Art installation and emerging documentary exploring ecological interdependence.

Don't Look Up (2021): https://www.imdb.com/title/tt11286314/
A dark satire that exaggerates real-world behaviors to expose how institutions and cultures respond to existential threats including climate change.

Books (Novels)

Markley, S. (2023). *The deluge*. Simon & Schuster.

An epic novel of intersecting lives amid climate collapse.

Kingsolver, B. (2013). *Flight behavior*. Harper Perennial.

A tale of butterfly migration and ecological disruption in Appalachia.

Stanley Robinson, K. (2021). *The ministry for the future: A novel.* Orbit.

A visionary narrative of policy, geoengineering, and global resistance.

REFERENCES

Ainslie, G. (2001). *Breakdown of will*. Cambridge University Press. https://doi.org/10.1017/cbo9781139164191

Akómoláfé, B. (2025, June 15). A slower urgency. *Writings*. https://www.bayoakomolafe.net/post/a-slower-urgency

Akómoláfé, B. (n.d.). *A slower urgency*. Bayo Akómoláfé. https://www.bayoakomolafe.net/post/a-slower-urgency

Andersson, J. J. (2019). Carbon taxes and CO_2 emissions: Sweden as a case study. *American Economic Journal: Economic Policy*, *11*(4), 1–30. https://doi.org/10.1257/pol.20170144

Arkes, H. R., & Blumer, C. (1985). The psychology of sunk costs. *Organizational Behavior and Human Decision Processes*, *35*(1), 124–140. https://doi.org/10.1016/0749-5978(85)90049-4

Arnsten, A. F. T. (2009). Stress signalling pathways that impair prefrontal cortex structure and function. *Nature Reviews Neuroscience*, *10*(6), 410–422. https://doi.org/10.1038/nrn2648

Arrien, A. (2012, September 27). Closure part—*The miracle bridge* [Transcript]. Values in World Thought, Mount Madonna School.

Ayalon, L., & Roy, S. (2022). A scoping review of research on older people and intergenerational relations in the context of climate change. *The Gerontologist, 63(5), 945–956.*

Aziz, A. (2024, October 1). *THE HERDS: A global act of theater for climate action*. Conspiracy of Love. https://www.conspiracyoflove.co/inspiration-articles/the-herds-a-global-act-of-theater-for-climate-action

Barrett, L. F. (2017). *How emotions are made: The secret life of the brain.* Houghton Mifflin Harcourt.

Barrett, L. F. (2024, August 8). The "fight or flight" idea misses the beauty of what the brain really does. *Scientific American.* https://www.scientificamerican.com/article/simplistic-fight-or-flight-idea-undervalues-the-brains-predictive-powers/

Baumeister, R. F., Bratslavsky, E., Finkenauer, C., & Vohs, K. D. (2001). Bad is stronger than good. *Review of General Psychology, 5*(4), 323–370. https://doi.org/10.1037//1089-2680.5.4.323

Beck, J. S. (2011). *Cognitive behavior therapy: Basics and beyond* (2nd ed.). Guilford Press.

Begum, N., Ali, A., Nazeef, N. M., & Gul, H. (2025). Cognitive biases and climate change decision-making: Understanding the role of optimism and anchoring in environmental choices. *The Critical Review of Social Sciences Studies, 3*(1), 2826–2838. https://doi.org/10.59075/btsmeb18

Benson, B. (2016, September 1). Cognitive bias cheat sheet. *Medium.* https://medium.com/thinking-is-hard/cognitive-bias-cheat-sheet-55a472476b18

Berman, M. G., Jonides, J., & Kaplan, S. (2008). The cognitive benefits of interacting with nature. *Psychological Science, 19*(12), 1207–1212. https://doi.org/10.1111/j.1467-9280.2008.02225.x

Bodansky, D. (2001). The history of the global climate change regime. In U. Luterbacher & D. F. Sprinz (Eds.), *International relations and global climate change* (pp. 23–40). MIT Press.

Boneva, T., Andre, P., Chopra, F., & Falk, A. (2024). Globally representative evidence on the actual and perceived support for climate action. *Nature Climate Change, 14*, 253–259. https://doi.org/10.1038/s41558-024-01925-3

Boyd, R. (2019, September 23). *The people of the Isle de Jean Charles are Louisiana's first climate refugees—they won't be the last.* Natural Resources Defense Council. https://www.nrdc.org/stories/people-isle-jean-charles-are-louisianas-first-climate-refugees-they-wont-be-last

Boyd, S. (2023, February 16). Constructive uncertainty. *Sunsama.* https://www.sunsama.com/blog/constructive-uncertainty

Brave Heart, M. Y. H. (1998). The return to the sacred path: Healing the historical trauma and historical unresolved grief response among the Lakota through a psychoeducational group intervention. *Smith College Studies in Social Work, 68*(3), 287–305.

Brigantine Living. (2025, March 28). *Turning trash into treasure: Meet Brigantine's local artist, Eduardo Jimenéz*. https://www.brigantineliving.online/blog/turning-beach-trash-into-art

brown, a. m. (2017). *Emergent strategy: Shaping change, changing worlds*. AK Press.

brown, a. m. (Ed.). (2019). *Pleasure activism: The politics of feeling good*. AK Press.

brown, a. m. (2020). *We will not cancel us: And other dreams of transformative justice*. AK Press.

Brulle, R. J. (2018). The climate lobby: A sectoral analysis of lobbying spending on climate change in the United States, 2000 to 2016. *Climatic Change*, *149*(3–4), 289–303. https://doi.org/10.1007/s10584-018-2241-z

Bullard, R. D. (Ed.). (1993). *Confronting environmental racism: Voices from the grassroots*. South End Press.

Bullard, R. D. (2000). *Dumping in Dixie: Race, class, and environmental quality* (3rd ed.). Westview Press.

Bullard, R. D. (Ed.). (2005). *The quest for environmental justice: Human rights and the politics of pollution*. Sierra Club Books.

Butler, L. J., Scammell, M. K., & Benson, E. B. (2016). The Flint, Michigan water crisis: A case study in regulatory failure and environmental injustice. *Environmental Justice*, *9*(4), 93–97. https://doi.org/10.1089/env.2016.0014

Camden FireWorks. (n.d.). *Seaside salvage*. https://www.camdenfireworks.org/seaside-salvage

Cantor, M. (2025, March 6). What is this era of calamity we're in? Some say "polycrisis" captures it. *The Guardian*. https://www.theguardian.com/us-news/2025/mar/06/polycrisis-disasters-politics

Carson, R. (1962). *Silent spring*. Houghton Mifflin.

Caverni, J.-P., Fabre, J.-M., & González, M. (Eds.). (1990). *Advances in psychology: Vol. 68. Cognitive biases: Their contribution for understanding human cognitive processes*. Elsevier/North-Holland. https://doi.org/10.1016/S0166-4115(08)61311-4

Center for Climate and Energy Solutions. (n.d.). *California Cap and Trade*. https://www.c2es.org/content/california-cap-and-trade/

Centola, D. (2018). *How behavior spreads: The science of complex contagions*. Princeton University Press.

Centola, D., Becker, J., Brackbill, D., & Baronchelli, A. (2018, June 8). Experimental evidence for tipping points in social convention. *Science*, *360*(6393), 1116–1119. https://doi.org/10.1126/science.aas8827

Chambers, R., Gullone, E., & Allen, N. B. (2009). Mindful emotion regulation: An integrative review. *Clinical Psychology Review*, *29*(6), 560–572. https://doi.org/10.1016/j.cpr.2009.06.005

Clayton, S. (2020). Climate anxiety: Psychological responses to climate change. *Journal of Anxiety Disorders*, *74*, Article 102263. https://doi.org/10.1016/j.janxdis.2020.102263

Clean Air Act of 1963, 42 U.S.C. §§ 7401–7671q (1963).

Climate Change Coaches. (2024). *Holding back climate progress: Sustainability's critical skills gap*. https://climatechangecoaches.com/sustainabilitys-new-skills-research/

Climate Critical. (2023). Climate Burnout Report. https://www.climatecritical.earth/report

Climate Justice Alliance. (n.d.). *The Principles of Environmental Justice*. https://climatejusticealliance.org/ej-principles/

Columbia University European Institute. (2019, January 18). Adam Tooze Named one of *Foreign Policy*'s Top 100 Global Thinkers for 2019. https://europe.columbia.edu/news/adam-tooze-named-one-foreign-policys-top-100-global-thinkers-2019

Coren, M. J. (2025, April 29). Writing to the future is one of the most powerful climate actions you can take. *The Washington Post*. https://www.washingtonpost.com/climate-environment/2025/04/29/climate-action-letter-to-the-future/

Cosentino, M., Gal-Oz, R., & Safer, D. L. (2025). Community-based resilience: The influence of collective efficacy and positive deviance on climate-related mental health. In E. Coren & H. Wang (Eds.), *Storytelling to accelerate climate solutions* (pp. 319–338). Springer. https://doi.org/10.1007/978-3-031-54790-4_15

Cosmides, L., & Tooby, J. (2000). Evolutionary psychology and the emotions. In M. Lewis & J. M. Haviland-Jones (Eds.), *Handbook of emotions* (2nd ed., pp. 91–115). Guilford Press.

Covey, S. R. (1989). *The seven habits of highly effective people: Powerful lessons in personal change*. Free Press.

Creswell, J. D. (2017). Mindfulness interventions. *Annual Review of Psychology*, *68*(1), 491–516. https://doi.org/10.1146/annurev-psych-042716-051139

Cunningham, S., & Ansorge, R. (2025, March 31). *Rest and recovery are critical for athletes of all ages from students to pros to older adults.* UCHealth Today. https://www.uchealth.org/today/rest-and-recovery-for-athletes-physiological-psychological-well-being/

Cunsolo, A. (2017). Climate change as the work of mourning. In A. Cunsolo & K. Landman (Eds.), *Mourning nature: Hope at the heart of ecological loss and grief* (pp. 169–191). McGill–Queen's University Press. https://doi.org/10.1515/9780773549357-011

Cunsolo, A., & Ellis, N. R. (2018). Ecological grief as a mental health response to climate change–related loss. *Nature Climate Change*, *8*(4), 275–281. https://doi.org/10.1038/s41558-018-0092-2

Cunsolo, A., & Landman, K. (Eds.). (2017). *Mourning nature: Hope at the heart of ecological loss and grief.* McGill-Queen's University Press.

Davenport, L. (2017). *Emotional resiliency in the era of climate change: A clinician's guide.* Jessica Kingsley.

DearTomorrow. (n.d.-a). *About us.* https://www.deartomorrow.org/about-us

DearTomorrow. (n.d.-b). *Write.* https://www.deartomorrow.org/write/

DNCO. (2022, November 24). *"Love is the running towards": Homage to London Fire Brigade.* https://dnco.com/thinking/love-is-the-running-towards-homage-to-london-fire-brigade

Endangered Species Act of 1973, 16 U.S.C. §§ 1531–1544 (1973).

Fabel, M., Ludwig, M., Flückiger, M., & Rainer, H. (2022). *The working power of youth: Political impacts of the "Fridays for Future" movement.* Working Paper No. 9742, Center for Economic Studies and Institute for Economic Research. https://www.econstor.eu/bitstream/10419/260872/1/cesifo1_wp9742.pdf

Fallon, B., & Greenberg, T. (2024, April 23). NJ man creates art from trash in effort to promote cleaner beaches at the Jersey Shore. *NBC10 Philadelphia.* https://www.nbcphiladelphia.com/news/local/nj-man-creates-art-from-trash-in-effort-to-promote-cleaner-beaches-at-the-jersey-shore/3839932/

Federal Water Pollution Control Act of 1972, 33 U.S.C. §§ 1251–1389 (1972).

Fisher, D. R., & Nasrin, S. (2021). Climate activism and its effects. *WIREs Climate Change*, *12*(1), e683. https://doi.org/10.1002/wcc.683

Franta, B. (2021). Early oil industry knowledge of CO_2 and global warming. *Nature Climate Change*, *8*, 1024–1025. https://doi.org/10.1038/s41558-018-0349-9

Fuller, R., Landrigan, P. J., Balakrishnan, K., Bathan, G., Bose-O'Reilly, S., Brauer, M., Caravanos, J., Chiles, T., Cohen, A., Corra, L., Cropper, M., Ferraro, G., Hanna, J., Hanrahan, D., Hu, H., Hunter, D., Janata, G., Kupka, R., Lanphear, B., . . . Yan, C. (2022). Pollution and health: A progress update. *The Lancet Planetary Health, 6*(6), e535–e547. https://doi.org/10.1016/S2542-5196(22)00090-0

Geisel, T. S. (1971). *The Lorax* [Dr. Seuss]. Random House.

Gifford, R. (2011). The dragons of inaction: Psychological barriers that limit climate change mitigation and adaptation. *American Psychologist*, *66*(4), 290–302. https://doi.org/10.1037/a0023566

Goldberg, M. H., van der Linden, S., Maibach, E., & Leiserowitz, A. (2019). Discussing global warming leads to greater acceptance of climate science. *Proceedings of the National Academy of Sciences*, *116*(30), 14804–14805. https://doi.org/10.1073/pnas.1906589116

Goodgoodgood. (2025, December 15). *Life-size animal puppet herd travels from the Congo Basin to a melting glacier in the Arctic*. https://www.goodgoodgood.co/articles/the-herds-puppets

Government Offices of Sweden. (2024). *Sweden's carbon tax*. https://www.government.se/government-policy/taxes-and-tariffs/swedens-carbon-tax/

Gross, J. J. (2015). Emotion regulation: Current status and future prospects. *Psychological Inquiry*, *26*(1), 1–26. https://doi.org/10.1080/1047840x.2014.940781

Gross, J. J., & Levenson, R. W. (1997). Hiding feelings: The acute effects of inhibiting negative and positive emotion. *Journal of Abnormal Psychology*, *106*(1), 95–103. https://doi.org/10.1037//0021-843x.106.1.95

Gyekye, K. (1995). *An essay on African philosophical thought: The Akan conceptual scheme* (Rev. ed.). Temple University Press.

Hanson, R. (2013). *Hardwiring happiness: The new brain science of contentment, calm, and confidence*. Harmony Books.

Harlan, S. L., Pellow, D. N., Roberts, J. T., Bell, S. E., Holt, W. G., & Nagel, J. (2015). Climate justice and inequality. In R. E. Dunlap & R. J. Brulle (Eds.), *Climate change and society: Sociological perspectives* (pp. 127–163). Oxford Academic. https://doi.org/10.1093/acprof:oso/9780199356102.003.0005

Harper, D. (n.d.). Emergency. In *Online etymology dictionary*. Retrieved October 17, 2025, from https://www.etymonline.com/word/emergency

Hawk, M., Coulter, R. W. S., Egan, J. E., Fisk, S., Reuel Friedman, M., Tula, M., & Kinsky, S. (2017). Harm reduction principles for healthcare settings. *Harm Reduction Journal*, *14*(1), Article 70. https://doi.org/10.1186/s12954-017-0196-4

Hayes, K., Blashki, G., Wiseman, J., Burke, S., & Reifels, L. (2018). Climate change and mental health: Risks, impacts and priority actions. *International Journal of Mental Health Systems*, *12*, Article 28. https://doi.org/10.1186/s13033-018-0210-6

Hayhoe, K. (2021). *Saving us: A climate scientist's case for hope and healing in a divided world.* Atria/One Signal.

Hayhoe, K [@katharinehayhoe.com]. (2024, August 20). *One of the most frequent questions I'm asked as a climate scientist is, "what can we do about climate change?"* [Article attached]. Bluesky. https://web-cdn.bsky.app/profile/katharinehayhoe.com/post/3l26cfjqdss2c

Held v. State of Montana, 560 P.3d 1235 (Mont. 2024).

Hernández, D. (2016). Understanding "energy insecurity" and why it matters to health. *Social Science & Medicine*, *167*, 1–10. https://doi.org/10.1016/j.socscimed.2016.08.029

Hickman, C., Marks, E., Pihkala, P., Clayton, S., Lewandowski, R. E., Mayall, E. E., Wray, B., Mellor, C., & van Susteren, L. (2021). Climate anxiety in children and young people and their beliefs about government responses: A global survey. *The Lancet Planetary Health*, *5*(12), e863–e873. https://doi.org/10.1016/S2542-5196(21)00278-3

Hofmann, S. G., Asnaani, A., Vonk, I. J. J., Sawyer, A. T., & Fang, A. (2012). The efficacy of cognitive behavioral therapy: A review of meta-analyses. *Cognitive Therapy and Research*, *36*(5), 427–440. https://doi.org/10.1007/s10608-012-9476-1

Holt-Lunstad, J., Smith, T. B., & Layton, J. B. (2010). Social relationships and mortality risk: A meta-analytic review. *PLOS Medicine, 7*(7), e1000316. https://doi.org/10.1371/journal.pmed.1000316

Hopi Elders. (2000, June 8). *Hopi prophecy: We are the ones we've been waiting for.* Awakin.org. https://www.awakin.org/v2/read/view.php?tid=702

Hsu, A., Sheriff, G., Chakraborty, T., & Manya, D. (2021). Disproportionate exposure to urban heat island intensity across major U.S. cities. *Nature Communications*, *12*(1), Article 2721. https://doi.org/10.1038/s41467-021-22799-5

Human Rights Research Center. (2025, February 21). *Unregulated contaminants in U.S. drinking water: Public health challenges and equitable solutions*. https://www.humanrightsresearch.org/post/unregulated-contaminants-in-u-s-drinking-water-public-health-challenges-and-equitable-solutions

Indigenized Energy Initiative. (2022, May 19). *Northern Cheyenne tribe selects Indigenized Energy Initiative to guide the tribe in the development of a $4.1M solar initiative on the reservation*. PR Newswire/Indigenous Energy News. https://www.prnewswire.com/news-releases/northern-cheyenne-tribe-selects-indigenized-energy-initiative-to-guide-the-tribe-in-the-development-of-a-4-1m-solar-initiative-on-the-reservation-301551522.html

Institute for Transportation & Development Policy. (2024, March). *The BRT standard*. https://itdp.org/publication/the-brt-standard/

Intergovernmental Panel on Climate Change. (2021). *Climate change 2021: The physical science basis*. https://www.ipcc.ch/report/ar6/wg1/

Intergovernmental Panel on Climate Change. (2022). *Climate change 2022: Impacts, adaptation and vulnerability*. https://www.ipcc.ch/report/ar6/wg2/

Intergovernmental Panel on Climate Change. (2023). *AR6 synthesis report: Climate change 2023*. https://doi.org/10.59327/IPCC/AR6-9789291691647

Intergovernmental Science–Policy Platform on Biodiversity and Ecosystem Services. (2019). *The global assessment report on biodiversity and ecosystem services: Summary for policymakers*. https://files.ipbes.net/ipbes-web-prod-public-files/inline/files/ipbes_global_assessment_report_summary_for_policymakers.pdf

Internal Displacement Monitoring Centre. (2024). *2024 Global report on internal displacement 2024*. https://www.internal-displacement.org/global-report/grid2024/

International Energy Agency. (2022, October 27). *World energy outlook 2022*. https://www.iea.org/reports/world-energy-outlook-2022

International Energy Agency. (2023a, March 1). CO_2 emissions in 2023. https://www.iea.org/reports/co2-emissions-in-2023

International Energy Agency. (2023b, January 11). Renewables 2023: Analysis and forecasts to 2028. https://www.iea.org/reports/renewables-2023

Jerath, R., Edry, J. W., Barnes, V. A., & Jerath, V. (2006). Physiology of long pranayamic breathing: Neural respiratory elements may provide a mechanism that explains how slow deep breathing shifts the autonomic nervous system. *Medical Hypotheses*, *67*(3), 566–571. https://doi.org/10.1016/j.mehy.2006.02.042

Juliana v. United States, 947 F.3d 1159 (9th Cir. 2020).

Kabat-Zinn, J. (1990). *Full catastrophe living: Using the wisdom of your body and mind to face stress, pain, and illness*. Delacorte.

Kahneman, D. (2011). *Thinking, fast and slow*. Farrar, Straus and Giroux.

Kahneman, D., & Tversky, A. (1979). Prospect theory: An analysis of decision under risk. *Econometrica*, *47*(2), 263–292. https://doi.org/10.2307/1914185

Kelly, C., Maguire, S., & Hevey, D. (2024). A scoping review of the impact of eco-distress and coping with distress on the mental health experiences of climate scientists. *Frontiers in Psychology*, *15*, 1351428. https://doi.org/10.3389/fpsyg.2024.1351428

Kimmerer, R. W. (2013). *Braiding sweetgrass: Indigenous wisdom, scientific knowledge, and the teachings of plants*. Milkweed Editions.

La Rose v. His Majesty the King, No. T-1750-19 (Federal Court of Canada, October 1, 2019).

Larsen, J. T., & McGraw, A. P. (2011). Further evidence for mixed emotions. *Journal of Personality and Social Psychology*, *100*(6), 1095–1110. https://doi.org/10.1037/a0021846

Lawrance, E. L., Thompson, R., Newberry Le Vay, J., Page, L., & Jennings, N. (2022, August). The impact of climate change on mental health and emotional wellbeing: A narrative review of current evidence, and its implications. *International Review of Psychiatry*, *34*(5), 443–498. https://doi.org/10.1080/09540261.2022.2128725

Lear, L. (1997). *Rachel Carson: Witness for nature*. Houghton Mifflin.

Lebel, L., Paquin, V., Kenny, T.-A., Fletcher, C., Nadeau, L., Chachamovich, E., & Lemire, M. (2022). Climate change and Indigenous mental health in the Circumpolar North: A systematic review to inform clinical practice. *Transcultural Psychiatry*, *59*(3), 312–336. https://doi.org/10.1177/13634615211066698

LeDoux, J. E. (1996). *The emotional brain: The mysterious underpinnings of emotional life*. Simon & Schuster.

Leghari v. Federation of Pakistan, W.P. No. 25501/2015 (Lahore High Court, September 4, 2015).

Leiserowitz, A., Maibach, E., Rosenthal, S., Kotcher, J., Goddard, E., Carman, J., Ballew, M., Verner, M., Marlon, J., Lee, S., Myers, T., Goldberg, M.,

Badullovich, N., & Thier, K. (2024, January 11). *Climate change in the American mind: Beliefs & attitudes, Fall 2023*. Yale Program on Climate Change Communication and George Mason University Center for Climate Change Communication. https://climatecommunication.yale.edu/publications/climate-change-in-the-american-mind-beliefs-attitudes-fall-2023/

Leiserowitz, A., Maibach, E., Rosenthal, S., Kotcher, J., Carman, J., Goddard, E., Myers, T., Verner, M., Marlon, J., Goldberg, M., Ettinger, J., Richards, E., Fine, J., & Thier, K. (2025). *Global Warming's six Americas, Fall 2024*. Yale Program on Climate Change Communication and George Mason University Center for Climate Change Communication. https://climatecommunication.yale.edu/publications/global-warmings-six-americas-fall-2024/

Leiserowitz, A., Verner, M., Goddard, E., Wood, E., Carman, J., Ordaz Reynoso, N., Thulin, E., Rosenthal, S., Marlon, J., & Buttermore, N. (2023). *International public opinion on climate change, 2023*. Yale Program on Climate Change Communication and Data for Good at Meta. https://climatecommunication.yale.edu/publications/international-public-opinion-on-climate-change-2023/

Lenton, T. M., Rockström, J., Gaffney, O., Rahmstorf, S., Richardson, K., Steffen, W., & Schellnhuber, H. J. (2019, November 27). Climate tipping points—Too risky to bet against. *Nature*, *575*(7784), 592–595. https://doi.org/10.1038/d41586-019-03595-0

Lertzman, R. (2023). Applying climate psychology principles and methods to the work of leading ecological transformation. *Field Actions Science Reports*, *25*, 88–91. https://journals.openedition.org/factsreports/7349

Lewandowski, R. E., Clayton, S. D., Olbrich, L., Sakshaug, J. W., Wray, B., Schwartz, S. E. O., Augustinavicius, J., Howe, P. D., Parnes, M., Wright, S., Carpenter, C., Wiśniowski, A., Ruiz, D. P., & Van Susteren, L. (2024). Climate emotions, thoughts, and plans among US adolescents and young adults: a cross-sectional descriptive survey and analysis by political party identification and self-reported exposure to severe weather events. *The Lancet Planetary Health*, *8*(11), E879–E893. https://doi.org/10.1016/S2542-5196(24)00229-8

Lindsey, R. (2025, May 21). *Climate change: Atmospheric carbon dioxide*. National Oceanic and Atmospheric Association. https://www.climate.

gov/news-features/understanding-climate/climate-change-atmospheric-carbon-dioxide

Lingo, K. J. (n.d.). *Bio*. Kaira Jewel Lingo. https://www.kairajewel.com/bio

Lingo, K. J. (2020, August 6). In times of crisis, draw upon the strength of peace. *Lion's Roar*. https://www.lionsroar.com/in-times-of-crisis-draw-upon-the-strength-of-peace/

Lingo, K. J. (2021). *We were made for these times: Ten lessons for moving through change, loss, and disruption*. Parallax Press.

Lorde, A. (1984). The uses of the erotic: The erotic as power. In A. Lorde, *Sister outsider: Essays and speeches* (pp. 110–114). Crossing Press.

Low-Income Solar Policy Guide. (2023). *Community powered solar*. https://www.lowincomesolar.org/community-powered-solar/

MacDonald, J. (1998). *The Arctic sky: Inuit astronomy, star lore, and legend*. Royal Ontario Museum.

Machado de Oliveira, V. (2021). *Hospicing modernity: Facing humanity's wrongs and the implications for social activism*. North Atlantic Books.

Madhusoodanan, J. (2023, December 18). Beyond bias. *OpenMind Magazine*. https://www.openmindmag.org/articles/immutable

Mann, M. E. (2023, September 14). *Climate doomism disregards the science*. APS News. https://www.aps.org/apsnews/2023/09/climate-doomism-disregards-science

Marks, J. (2024, April 25). *Davis company farms without using soil: Gotham Greens uses 90% less water.* KCRA 3 News. https://www.kcra.com/article/davis-company-gotham-greens/60607289

Marshall, J. M., III (2001). *The Lakota way: Stories and lessons for living*. Viking Compass.

Martinez, C., Kilbury, L., & Martinez, J., White, C., Lutz, M., So, K., Petosa, K., McManus, A., & Christianson, A. (2023, December 5). These fossil fuel industry tactics are fueling Democratic backsliding. *American Progress*. https://www.americanprogress.org/article/these-fossil-fuel-industry-tactics-are-fueling-democratic-backsliding/

Mayer, F. S., & Frantz, C. M. (2004). The Connectedness to Nature scale: A measure of individuals' feeling in community with nature. *Journal of Environmental Psychology*, *24*(4), 503–515. https://doi.org/10.1016/j.jenvp.2004.10.001

McCall, R. (2019, September 19). #NoFutureNoChildren: Hundreds of young people pledge not to have children for the environment. *Newsweek*. https://www.newsweek.com/hundreds-people-pledge-no-children-climate-change-1460140

McEwen, B. S. (1998). Protective and damaging effects of stress mediators. *The New England Journal of Medicine*, *338*(3), 171–179. https://doi.org/10.1056/nejm199801153380307

McKibben, B. (2017, April 4). The question I get asked the most. *Resilience*. https://www.resilience.org/stories/2017-04-04/question-get-asked/

McKibben, B. (2020, April 30). How to combat climate depression. *The New Yorker*. https://www.newyorker.com/news/annals-of-a-warming-planet/how-to-combat-climate-depression

McLaren, K. (2023). *The language of emotions: What your feelings are trying to tell you* (Rev ed.). Sounds True.

McMullen, H., & Dow, K. (2022). Ringing the existential alarm: Exploring BirthStrike for climate. *Medical Anthropology*, *41*(6–7), 659–673. https://doi.org/10.1080/01459740.2022.2083510

Mejia, N. (2021, April). *Climate activist Xiye Bastida wants you to "get out there and do it."* Mission. https://www.missionmag.org/climate-activist-xiye-bastida-wants-you-to-get-out-there-and-do-it/

Menakem, R. (2017). *My grandmother's hands: Racialized trauma and the pathway to mending our hearts and bodies*. Central Recovery Press.

Mennin, D. S., & Fresco, D. M. (2014). Emotion regulation therapy. In J. J. Gross (Ed.), *Handbook of emotion regulation* (2nd ed., pp. 469–490). Guilford Press.

Mitchell, J. (1970). Big yellow taxi [Song]. On *Ladies of the Canyon*. Reprise Records.

Moench, M. (2023, November 30). New data shows a global surge in searches related to "climate anxiety." *Time*. https://time.com/6338759/climate-change-anxiety-google-search-trend/

Moser, S. C. (2020). The adaptive mind. In A. E. Johnson & K. K. Wilkinson (Eds.), *All we can save: Truth, courage, and solutions for the climate crisis* (pp. 270–278). One World.

Moss-Racusin, C. A., Dovidio, J. F., Brescoll, V. L., Graham, M. J., & Handelsman, J. (2012). Science faculty's subtle gender biases favor male students. *Proceedings of the National Academy of Sciences*, *109*(41), 16474–16479. https://doi.org/10.1073/pnas.1211286109

Mt. Pleasant, J. (2006). The science behind the Three Sisters mound system: An agronomic assessment of an Indigenous agricultural system in the Northeast. In J. E. Staller, R. H. Tykot, & B. F. Benz (Eds.), *Histories of maize: Multidisciplinary approaches to the prehistory, linguistics, biogeography, domestication, and evolution of maize* (pp. 529–537). Academic Press.

Moyer, J., Smith, A., Rui, Y., & Hayden, J. (2020). *Regenerative agriculture and the soil carbon solution* [White paper]. Rodale Institute. https://rodaleinstitute.org/wp-content/uploads/Rodale-Soil-Carbon-White-Paper_v11-compressed.pdf

Myers, D. G., & Twenge, J. M. (2019). *Social psychology* (13th ed.). McGraw-Hill Education.

National Academies of Sciences, Engineering, and Medicine. (2019). *Negative emissions technologies and reliable sequestration: A research agenda 2019*. https://www.nationalacademies.org/projects/DELS-BASCPR-16-01/publication/25259

National Aeronautics and Space Administration. (2024). *Scientific consensus*. https://climate.nasa.gov/scientific_consensus/

National Museum of the American Indian, Education Office. (2009.). *Haudenosaunee guide for educators*. Smithsonian Institution. https://americanindian.si.edu/sites/1/files/pdf/education/haudenosaunee guide.pdf?utm

Nelson, A. (2012). *Leaf litter talks with Dr. Robin Wall Kimmerer*. Biohabitats. https://www.biohabitats.com/newsletter/traditional-ecological-knowledge/leaf-litter-talks-with-dr-robin-wall-kimmerer/

Nesse, R. M. (2019). *Good reasons for bad feelings: Insights from the frontier of evolutionary psychiatry*. Dutton.

Nickerson, R. S. (1998). Confirmation bias: A ubiquitous phenomenon in many guises. *Review of General Psychology*, *2*(2), 175–220. https://doi.org/10.1037//1089-2680.2.2.175

O'Donoghue, T., & Rabin, M. (1999). Doing it now or later. *American Economic Review*, *89*(1), 103–124. https://doi.org/10.1257/aer.89.1.103

Ogden, P., Minton, K., & Pain, C. (2006). *Trauma and the body: A sensorimotor approach to psychotherapy*. W. W. Norton.

One Earth. (n.d.). *Climate hero: Xiye Bastida*. https://www.oneearth.org/climate-hero-xiye-bastida/

Otto, I. M., Donges, J. F., Cremades, R., Bhowmik, A., Hewitt, R. J., Lucht, W., . . . & Schellnhuber, H. J. (2020). Social tipping dynamics for stabilizing Earth's climate by 2050. *Proceedings of the National Academy of Sciences*, *117*(5), 2354–2365.

Palinkas, L. A., Hurlburt, M. S., Fernandez, C., De Leon, J., Yu, K., Salinas, E., Garcia, E., Johnston, J., Rahman, M. M., Silva, S. J., & McConnell, R. S. (2022). Vulnerable, resilient, or both? A qualitative study of adaptation resources and behaviors to heat waves and health outcomes of low-income residents of urban heat islands. *International Journal of Environmental Research and Public Health*, *19*(17), Article 11090. https://doi.org/10.3390/ijerph191711090

Pennebaker, J. W., & Chung, C. K. (2011). Expressive writing: Connections to physical and mental health. In H. S. Friedman (Ed.), *The Oxford handbook of health psychology* (pp. 417–437). Oxford University Press. https://doi.org/10.1093/oxfordhb/9780195342819.013.0018

Pihkala, P. (2019). Anxiety and the ecological crisis: An analysis of eco-anxiety and climate anxiety. *Sustainability*, *11*(19), 5244.

Pihkala, P. (2020a). Anxiety and the ecological crisis: An analysis of eco-anxiety and climate anxiety. *Sustainability*, *12*(19), Article 7836. https://doi.org/10.3390/su12197836

Pihkala, P. (2020b). Eco-anxiety and environmental education. *Sustainability*, *12*(23), 10149. https://doi.org/10.3390/su122310149

Pihkala, P. (2022). The process of eco-anxiety and ecological grief: A narrative review and a new proposal. *Sustainability*, *14*(24), 16628. https://doi.org/10.3390/su142416628

Pihkala, P. (2022). Toward a taxonomy of climate emotions. *Frontiers in Climate*, *3*, Article 863460. https://doi.org/10.3389/fclim.2021.738154

Pihkala, P. (2024). Ecological sorrow: Types of grief and loss in ecological grief. *Sustainability*, *16*(2), Article 849. https://doi.org/10.3390/su16020849

Pihkala, P., & Kamenetz, A. (2024). *A guide to climate emotions*. Climate Mental Health Network. https://www.researchgate.net/publication/380186405_A_Guide_to_Climate_Emotions

Porges, S. W. (2011). *The polyvagal theory: Neurophysiological foundations of emotions, attachment, communication, and self-regulation*. W. W. Norton.

Porges, S. W., & Dana, D. (2018). *Clinical applications of the polyvagal theory: The emergence of polyvagal-informed therapies*. W. W. Norton.

Potential Energy Coalition. (2023). *Later is too late to act on climate change.* https://potentialenergycoalition.org/later-is-too-late-global-report/

Powers, R. (2018). *The overstory.* W. W. Norton & Company.

Prather, M. J. (2025). Calibrating the tropospheric air and ozone mass. *AGU Advances*, 6, e2025AV001651. https://doi.org/10.1029/2025AV001651

Pregadio, F. (Ed.). (2008). *The encyclopedia of Taoism* (Vols. 1–2). Routledge.

Randall, R. (2009). Loss and climate change: The cost of parallel narratives. *Ecopsychology*, *1*(3), 118–129. https://doi.org/10.1089/eco.2009.0034

Ray, S. J. (2020). *A field guide to climate anxiety: How to keep your cool on a warming planet.* University of California Press.

Rittel, H. W. J., & Webber, M. M. (1973). Dilemmas in a general theory of planning. *Policy Sciences*, *4*(2), 155–169. https://doi.org/10.1007/bf01405730

Roberts, M., Haami, B., Benton, R. A., Satterfield, T., Finucane, M. L., & Henare, M. (2004). Whakapapa as a Maori mental construct: Some implications for the debate over genetic modification of organisms. *The Contemporary Pacific*, *16*(1), 1–28.

Rozin, P., & Royzman, E. B. (2001). Negativity bias, negativity dominance, and contagion. *Personality and Social Psychology Review*, *5*(4), 296–320. https://doi.org/10.1207/s15327957pspr0504_2

Runte, A. (2010). *National parks: The American experience* (4th ed.). Taylor Trade Publishing.

Rush, C. (2025, March 24). *A decadelong climate lawsuit saw plaintiffs grow from childhood to adulthood. Now, it's over.* Associated Press. https://apnews.com/article/b90f095046663da0e693b6f2fc46e71e

Salovey, P., & Mayer, J. D. (1990). Emotional intelligence. *Imagination, Cognition and Personality*, *9*(3), 185–211. https://doi.org/10.2190/DUGG-P24E-52WK-6CDG

St. Martin, V. (2024, June 1). Who wants to have children in a warming world? *Wired.* https://www.wired.com/story/babies-warming-world-jade-sasser-climate-change-anxiety/

Sawin, E. (2024). *Multisolving: Creating systems change in a fractured world.* Island Press.

Schauer, M., & Elbert, T. (2010). Dissociation following traumatic stress. *Journal of Psychology*, *218*(2), 109–127. https://doi.org/10.1027/0044-3409/a000018

Schiffman, R. (2022, April 28). For Gen Z, climate change is a heavy emotional burden. *Yale Environment 360*. https://e360.yale.edu/features/for-gen-z-climate-change-is-a-heavy-emotional-burden

Schwartz, S. E. O., Benoit, L., Clayton, S., Parnes, M. F., Swenson, L., & Lowe, S. R. (2022). Climate change anxiety and mental health: Environmental activism as buffer. *Current Psychology*, *42*, 16708–16721. https://doi.org/10.1007/s12144-022-02735-6

Secretariat of the Convention on Biological Diversity. (2020). *Global Biodiversity Outlook 5*. https://www.cbd.int/gbo/gbo5/publication/gbo-5-en.pdf

Seydi, S. T. (2025). *Assessment of the January 2025 Los Angeles County wildfires: A multi-modal analysis of impact, response, and population exposure*. arXiv preprint. https://arxiv.org/abs/2501.17880

Sharot, T. (2011). The optimism bias. *Current Biology*, *21*(23), R941–R945. https://doi.org/10.1016/j.cub.2011.10.030

Shrum, T. R. (2021). The salience of future impacts and the willingness to pay for climate change mitigation: An experiment in intergenerational framing. *Climatic Change*, *164*(3–4), 40. https://doi.org/10.1007/s10584-021-03002-6

Siclari, F., Baird, B., Perogamvros, L., Bernardi, G., LaRocque, J. J., Riedner, B., Boly, M., Postle, B. R., & Tononi, G. (2017). The neural correlates of dreaming. *Nature Neuroscience*, *20*(6), 872–878. https://doi.org/10.1038/nn.4545

Siegel, D. J. (1999). *The developing mind: Toward a neurobiology of interpersonal experience*. Guilford Press.

Siegel, D. J. (2012). *The developing mind: How relationships and the brain interact to shape who we are* (2nd ed.). Guilford Press.

Slovic, P. (2007). "If I look at the mass I will never act": Psychic numbing and genocide. *Judgment and Decision Making*, *2*(2), 79–95. https://doi.org/10.1017/s1930297500000061

Smith, B. W., Dalen, J., Wiggins, K., Tooley, E., Christopher, P., & Bernard, J. (2008). The Brief Resilience Scale: Assessing the ability to bounce back. *International Journal of Behavioral Medicine*, *15*(3), 194–200. https://doi.org/10.1080/10705500802222972

Stoknes, P. E. (2015). *What we think about when we try not to think about global warming: Toward a new psychology of climate action*. Chelsea Green Publishing.

Sunrise Movement. (n.d.). *About Sunrise*. https://www.sunrisemovement.org/about/

Supran, G., & Oreskes, N. (2017). Assessing ExxonMobil's climate change communications (1977–2014). *Environmental Research Letters*, *12*(8), Article 084019. https://doi.org/10.1088/1748-9326/aa815f

Supran, G., & Oreskes, N. (2021). Rhetoric and frame analysis of ExxonMobil's climate change communications. *One Earth*, *4*(5), P696–P719. https://doi.org/10.1016/j.oneear.2021.04.014

Swim, J. K., Geiger, N., & Swim, J. L. (2022). OK Boomer: A decade of generational differences in feelings about climate change. *Global Environmental Change*, *73*, 102478.

Syropoulos, S., Markowitz, E. M., & Shrum, T. A. (2023). A letter to future generations: Examining the effectiveness of an intergenerational framing intervention. *Journal of Environmental Psychology*, *90*, Article 102074.

Tannen, D. (1998). *The argument culture: Moving from debate to dialogue*. Random House.

Taylor, D. E. (2014). *Toxic communities: Environmental racism, industrial pollution, and residential mobility*. NYU Press.

Terrell, K. A., & St. Julien, G. (2022). Air pollution is linked to higher cancer rates among Black or impoverished communities in Louisiana. *Environmental Research Letters*, *17*(1), Article 014033. https://doi.org/10.1088/1748-9326/ac4360

Thiem, H., & Lindsey, R. (2024, November 7). *Hurricane Helene's extreme rainfall and catastrophic flooding*. National Oceanic and Atmospheric Administration. https://www.climate.gov/news-features/event-tracker/hurricane-helenes-extreme-rainfall-and-catastrophic-inland-flooding

Thomas, K., Hardy, R. D., Lazrus, H., Mendez, M., Orlove, B., Rivera-Collazo, I., Roberts, J. T., Rockman, M., Warner, B. P., & Winthrop, R. (2019). Explaining differential vulnerability to climate change: A social science review. *Wiley Interdisciplinary Reviews: Climate Change*, *10*(2), e565. https://doi.org/10.1002/wcc.565

Torres, P. (2023, November 6). *Find the joy in environmental work; it's there*. Yale School of the Environment. https://environment.yale.edu/news/article/find-joy-environmental-work-its-there

Trouet, V., Babst, F., & Meko, M. (2018). Recent enhanced high-summer North Atlantic Jet variability emerges from three-century context.

Nature Communications, *9*, 180. https://doi.org/10.1038/s41467-017-02699-3

Tversky, A., & Kahneman, D. (1974, September 27). Judgment under uncertainty: Heuristics and biases. *Science*, *185*(4157), 1124–1131. https://doi.org/10.1126/science.185.4157.1124

Ulrich, R. S., Simons, R. F., Losito, B. D., Fiorito, E., Miles, M. A., & Zelson, M. (1991). Stress recovery during exposure to natural and urban environments. *Journal of Environmental Psychology*, *11*(3), 201–230.

United Nations. (1992). United Nations Framework Convention on Climate Change. https://unfccc.int/resource/docs/convkp/conveng.pdf

United Nations Environment Programme. (2023). *Emissions gap report 2023: Broken record – Temperatures hit new highs, yet world fails to cut emissions (again)*. https://doi.org/10.59117/20.500.11822/43922

UN Development Programme & University of Oxford. (2024, June 20). *80 percent of people globally want stronger climate action by governments according to UN Development Programme survey* [Press release]. https://www.undp.org/press-releases/80-percent-people-globally-want-stronger-climate-action-governments-according-un-development-programme-survey

UN Environment Programme. (1987). *Montreal Protocol on Substances That Deplete the Ozone Layer*. https://ozone.unep.org/treaties/montreal-protocol

UN Office for Disaster Risk Reduction. (2022). *Global assessment report on disaster risk reduction 2022*. https://www.undrr.org/gar/gar2022-our-world-risk-gar#container-downloads

United Nations. (1997). *Kyoto Protocol to the United Nations Framework Convention on Climate Change*. https://unfccc.int/resource/docs/convkp/kpeng.pdf

United Nations. (2015). *Paris Agreement*. https://unfccc.int/sites/default/files/english_paris_agreement.pdf

Uppalapati, S., Ballew, M., Campbell, E., Kotcher, J., Rosenthal, S., Leiserowitz, A., & Maibach, E. (2023). *The prevalence of climate change psychological distress among American adults. Yale University and George Mason University*. Yale Program on Climate Change Communication.

Urgenda Foundation v. The Netherlands, No. 19/00135 (Supreme Court of the Netherlands, December 20, 2019).

van der Kolk, B. A. (2014). *The body keeps the score: Brain, mind, and body in the healing of trauma*. Viking.

van der Linden, S., Maibach, E., & Leiserowitz, A. (2015). Improving public engagement with climate change: Five "best practice" insights from psychological science. *Perspectives on Psychological Science*, *10*(6), 758–763.

Van Susteren, L. (2018). The psychological impacts of the climate crisis: A call to action. *BJPsych International*, *15*(2), 25–26. https://doi.org/10.1192/bji.2017.40

Većkalov, B., Zarzeczna, N., McPhetres, J., van Harreveld, F., & Rutjens, B. T. (2024). Psychological distance to science as a predictor of science skepticism across domains. *Personality and Social Psychology Bulletin*, *50*(1), 18–37. https://doi.org/10.1177/01461672221118184

Vlasceanu, M., Doell, K. C., Bak-Coleman, J. B., Todorova, B., Berkebile-Weinberg, M. M., Grayson, S. J., Patel, Y., Goldwert, D., Pei, Y., Chakroff, A., Pronizius, E., van den Broek, K. L., Vlasceanu, D., Constantino, S., Morais, M. J., Schumann, P., Rathje, S., Fang, K., Aglioti, S. M., . . . Van Bavel, J. J. (2024). Addressing climate change with behavioral science: A global intervention tournament in 63 countries. *Science Advances*, *10*(6), Article eadj5778. https://doi.org/10.1126/sciadv.adj5778

Walker, A. (2006). *We are the ones we have been waiting for: Inner light in a time of darkness*. New Press.

Weather.com Meteorologists. (2024, May 17). *Severe outbreak spawned over 165 tornadoes, including Barnsdall, Oklahoma EF4*. The Weather Channel. https://weather.com/storms/tornado/news/2024-05-02-severe-weather-tornado-forecast-plains-midwest-south-may

Weinstein, E. (2021, Spring). Hope is an action verb. *Kenyon College Alumni Magazine*. https://bulletin.kenyon.edu/article/hope-is-an-action-verb/

Weinstein, N., Przybylski, A. K., & Ryan, R. M. (2009). Can nature make us more caring? Effects of immersion in nature on intrinsic aspirations and generosity. *Personality and Social Psychology Bulletin*, *35*(10), 1315–1329.

Widdicombe, L. (2021, December 27). How Xiye Bastida became a leader in the climate fight. *Vogue*. https://www.vogue.com/article/xiye-bastida-climate-activist-profile

Williams, T. T. (2002). *Red: Passion and patience in the desert*. Vintage Books.

Winters, J. (2025, January 3). "We have been heard": Montana youth score a major climate victory in court. *Grist.* https://grist.org/regulation/held-v-montana-youth-climate-lawsuit-supreme-court-decision/

Wong, P. T. P. (2012). Toward a dual-systems model of what makes life worth living. In P. T. P. Wong (Ed.), *The human quest for meaning: Theories, research, and applications* (2nd ed., pp. 3–22). Routledge.

World Bank. (2021, September 13). *Climate change could force 216 million people to migrate within their own countries by 2050.* https://www.worldbank.org/en/news/press-release/2021/09/13/climate-change-could-force-216-million-people-to-migrate-within-their-own-countries-by-2050

World Bank. (n.d.). *Country climate and development reports.* https://www.worldbank.org/en/publication/country-climate-development-reports

World Economic Forum. (2025). *The global risks report 2025* (20th ed.). https://reports.weforum.org/docs/WEF_Global_Risks_Report_2025.pdf

World Wildlife Fund. (2022, October 13). *69% average decline in wildlife populations since 1970, says new WWF report* [Press release]. https://www.worldwildlife.org/news/press-releases/69-average-decline-in-wildlife-populations-since-1970-says-new-wwf-report/

Wray, B. (2020, August 5). Why activism isn't *really* the cure for eco-anxiety and eco-grief. *Unthinkable Earth.* https://unthinkable.substack.com/p/why-activism-isnt-really-the-cure

Wray, B. (2022, May 5). Should you have kids in the midst of the climate crisis? *Chatelaine.* https://chatelaine.com/living/having-kids-climate-crisis/

Wray, B., & Drury, I. (2023, June 29). Please stop saying my generation will save the world. *Unthinkable.* https://unthinkable.substack.com/p/please-stop-saying-my-generation

Wynes, S., & Nicholas, K. A. (2017). The climate mitigation gap: Education and government recommendations miss the most effective individual actions. *Environmental Research Letters, 12*(7), 074024. https://doi.org/10.1088/1748-9326/aa7541

Yalom, I. D. (1980). *Existential psychotherapy.* Basic Books.

Youth v. Government of Mexico, No. 1854/2019 (District Court in Administrative Matters, Mexico, September 2, 2020).

Zero Hour. (2025). *Tackling the climate & nature crisis together*. https://www.zerohour.uk/climate-and-nature-bill

Zomorodi, M. (Host). (2020, May 22). How are young people making the choice to fight climate change? [Radio broadcast episode]. In *TED Radio Hour*. NPR. https://www.npr.org/transcripts/860168455

Zuabi, A. N. (2024, April). *The Herds, a vast act of theater to spark climate action* [Video]. TED Conferences. https://www.youtube.com/watch?v=30a4OlkenNU

INDEX

ABOUT THE AUTHOR

Leslie Davenport, MS, MA, is an internationally recognized climate psychology speaker, consultant, and therapist who helps people navigate the emotional landscape of living in a rapidly warming world. She brings decades of experience as an educator to the emerging field of climate psychology, where she integrates insights from psychology, science, and social justice to support both resilience and collective action. She is the author of five books, including *Emotional Resiliency in the Era of Climate Change* (2017), for mental health professionals, as well as *All the Feelings Under the Sun: How to Deal With Climate Change* (2021) and *What to Do When Climate Change Scares You: A Kid's Guide to Dealing with Climate Change Stress* (2024), both of which were written to guide young people through their climate emotions. She writes the "Ask a Climate Therapist" column for award-winning environmental news outlet Grist. Widely recognized for making complex ideas accessible, Leslie offers practical, compassionate tools that empower people of all ages to engage with courage and join in collective transformation.

In addition to her writing, Leslie advises organizations such as the Post Carbon Institute (https://www.postcarbon.org/), the Climate Mental Health Network (https://www.climatementalhealth.net/), and

One Resilient Earth (https://oneresilientearth.org/). She also leads the Climate Psychology Certificate program at the California Institute of Integral Studies, the only university-based program of its kind in the United States. Through public speaking, teaching, and consulting, Leslie helps transform climate distress into meaningful pathways for healing and change. Visit https://www.lesliedavenport.com, and follow her at https://www.instagram.com/ldavenportauthor.